The TRAVELLING NATURALISTS

The TRAVELLING NATURALISTS

Clare Lloyd

University of Washington Press
Seattle

For Paul, with love and thanks

Copyright © 1985 by Clare Lloyd
Maps © 1985 by Ulrica Lloyd

Published in the United States in 1985 by the
University of Washington Press in cooperation with
Croom Helm Ltd., Kent, England

All rights reserved. No part of this publication
may be reproduced or transmittted in any form or
by any means, electronic or mechanical, including
photocopy, recording, or any information storage
or retrieval system, without permission in writing
from the publisher.

Library of Congress Cataloguing in Publication Data

Lloyd, Clare
  The travelling naturalists.

  Bibliography: p. 150
  Includes index.
  1. Naturalists – Great Britain – Biography. I. Title.
QH26.L597 1985 508′.0922 [B] 85-11266
ISBN 0-295-96304-2

Printed and bound in Great Britain

CONTENTS

List of Maps	6
List of Black and White Illustrations	6
List of Colour Plates	7
Foreword	9
1. The Golden Age of the Naturalist	11
2. On Being a Travelling Naturalist	19
3. The Wanderer of Walton Hall	27
4. The Fate of Sir John Franklin	41
5. Eleven Years on the Amazon	61
6. The Source of the Nile?	75
7. Rambles in the Andes	93
8. To the Yenesei and Petchora	107
9. Thousands of Feet Up, Hundreds of Fathoms Down	121
10. Fishes and Fetish	135
11. Epilogue: the End of an Era	147
Select Bibliography	150
Postscript	152
Index	153

Maps

1 Places visited by Charles Waterton in Guiana (modern Guyana) and other parts of South America, 1804–24 — 26
2 The Canadian Arctic explored by John Franklin, Leopold McClintock and others, 1845–59, showing the position of the Northwest Passage — 40
3 Henry Bates's journeys in Brazil, 1848–59 — 60
4 Speke's journeys in East Africa, 1854–63 — 74
5 Howard Saunders's journey across the Andes, 1861–2 — 92
6 Henry Seebohm's journeys to west and east Siberia, 1875 and 1877 — 106
7 Green's journeys to Australia and New Zealand, 1882, and the Canadian Selkirks, 1888 — 120
8 West Africa, showing places visited by Mary Kingsley, 1893–5 — 134

Black & White Illustrations

1.1 Tribesmen from near Lake Tanganyika — 11
1.2 Grant's gazelle — 13
1.3 A Yurak tribesman, from eastern Siberia — 17
1.4 Butterflies of the genus *Heliconius* — 17
1.5 Hippopotamuses enjoying a bath — 18
2.1 Explorers in East Africa: Burton and Speke — 19
2.2 Isabella Bird in her 'Hawaiian riding dress' — 21
2.3 An alligator attacks Henry Bates's camp — 23
2.4 A mosquito veil used in Siberia — 25
3.1 Walton Hall, the Waterton family home — 27
3.2 Horned and crested screamers — 29
3.3 A pair of peccaries — 30
3.4 A sloth in the wild — 31
3.5 Waterton riding a caiman — 36
3.6 Waterton's Nondescript — 37
4.1 Sir Roderick Impey Murchison — 41
4.2 Admiral Sir Leopold McClintock — 43
4.3 A Greenlander's supper appropriated by a bear — 49
4.4 The *Fox* aground temporarily off Buchan Island — 51
4.5 McClintock reconnoitres the Bellot Strait in a rowing boat — 52
4.6 Ivory gulls — 58
4.7 Admiralty House, Bermuda — 59
5.1 Scarlet-faced monkeys and the monk saki — 61
5.2 Bates hunting toucans — 64
5.3 Bird-eating spider — 67
5.4 South American anteater — 68
5.5 Turtle — 70
6.1 John Hanning Speke — 75
6.2 Speke is chased by buffaloes — 84
6.3 Sitatunga waterbucks, a new species of antelope — 86
6.4 Speke presents his 'spoils' to King Rumanika — 87
6.5 The goatsucker, a species of nightjar — 89
7.1 Hawk-moths — 93
7.2 Capybara — 99
7.3 Manatee — 100
7.4 Hoatzin — 102
7.5 Hoopoe — 104
8.1 Henry Seebohm — 107
8.2 Grey plovers — 111
8.3 Seebohm travelling by rosposki — 114

8.4	Seebohm learning how to move on snow-shoes	115	9.5 Henry Swanzy hunting	131
			10.1 Mary Kingsley	135
8.5	The *Thames* grounded on submerged banks	116	10.2 Hunting a gorilla	142
			10.3 Leopard	144
9.1	Wallabies	121	10.4 *Ctenopoma kingsleyae*, discovered by Mary Kingsley in West Africa	146
9.2	The *Lord Bandon*'s dredging crew, 1885	128	11.1 Charles Darwin	147
9.3	Starfish and sea urchins	129		
9.4	Caribou in the Selkirk mountains	130		

Colour Plates

(Between pages 80 and 81)

C1. John James Audubon, one of the best-known of nineteenth-century bird artists, was an American and also a traveller. These five auks probably were drawn from specimens he saw in London during 1837. On the left are an adult and juvenile ancient murrelet, a least and crested auklet are swimming in the sea and on the right is a rhinoceros auklet. (Courtesy of Alexander Library, EGI, Oxford)

C2. The Arctic hare is camouflaged throughout the year. In summer it looks like any other hare, but in winter it grows a pure white coat. McClintock found few hares on the islands of Arctic Canada. In one eleven-month period of hunting for his crew's larder he shot only nine Arctic hares.

C3 & 4. These beautiful hand-coloured engravings of a common bee-eater and a teal appeared in a five-volume publication *A History of the Birds of Europe*. Its author was Henry Dresser, a worldwide traveller and collector of bird skins. When drawing one of these illustrations, Dresser used Henry Saunders's bee-eater skins, from birds Saunders had acquired in Spain in 1868. (Courtesy of Alexander Library, EGI, Oxford)

C5 Charles Waterton was a great admirer of Alexander Wilson who drew the picture for this engraving of a wild turkey cock. In 1824, eleven years after Wilson's death, Waterton made a special journey to Philadelphia where Wilson's book had been published. (Courtesy of Alexander Library, EGI, Oxford)

C6. One subspecies of this Cuvier's toucan lives between the Tapajos and Tocatins Rivers, tributaries of the River Amazon visited by Henry Bates. Both he and Howard Saunders saw many toucans in the Amazonian and Andean rain forests, but the First Cuvier's toucan to reach Europe was sent to London Zoo only in 1871. (Courtesy of Alexander Library, EGI, Oxford)

C7 & 8. Illustrations like these helped Philip Gosse's books on 'the wonders of the deep sea' to sell to the general public. Gosse also published more serious books on marine biology including one on the sea anemones and corals of Great Britain. (Courtesy of the Bodleian Library, Oxford: (18943 d.12 Plates VI and IX))

C9. Philip Gosse called this 'The Ancient Wrasse.' It is a chromolithograph made from an engraving by Gosse, and it appeared as a frontispiece to his book *The Aquarium* which was a bestseller in the 1850s. (Courtesy of the Bodleian Library, Oxford: (189.c58))

C10. Sir Richard Burton, a portrait by Lord Leighton painted in 1876 and now in the National Portrait Gallery. After his

African expeditions, Burton's life continued to be filled with 'tumultuous journeys' across three continents, and with quarrels and humiliations. One of his pastimes was to translate and publish Eastern erotica, including the famous *Thousand and One Nights*. (Courtesy of the National Portrait Gallery, London)

C11.   Thomas Huxley supported Darwin's controversial theories on evolution, and publicly defended them, earning himself the nickname of 'Darwin's Bulldog'. As a young man he spent six years surveying the coasts of Australia. On this expedition he became increasingly interested in marine biology and he continued as a naturalist after his return to England. (Courtesy of the National Portrait Gallery, London)

C12.   This is one of the only portraits of Charles Waterton because he detested sitting for an artist. George Ord, Waterton's friend from Philadelphia, bought the picture after Waterton's death and donated it to the Waterton Estate. It was subsequently purchased by the National Portrait Gallery.

C13.   The beetles in this lithograph are from the genera *Pelidonota* and *Plustiotis*, collected by Henry Bates in Brazil. He used his knowledge of South American entomology and his vast insect collection to contribute to a multi-volume work *Biologica Centrali-Americana*. (Courtesy of the Bodleian Library, Oxford: (19177 c1/18 Table 16)).

C14.   Howard Saunders first collected Bonelli's eagle chicks like these near Seville in 1868. He brought them back alive to London zoo. His skins from later visits to southern Spain were used by Henry Dresser when making the drawing for this engraving. (Courtesy of Alexander Library, EGI, Oxford)

C15.   Henry Seebohm was an expert on birds in the thrush family. This illustration of an American robin appeared in Seebohm's monograph on thrushes published after his death. (Courtesy of Alexander Library, EGI, Oxford)

C16.   As well as collecting bird skins, Henry Seebohm amassed a huge number of eggs. These are thrushes' eggs in a plate from his *Eggs of British Birds*, one of the first books to show eggs in colour. After his death, Seebohm's ornithological collection was given to the British Natural History Museum. (Courtesy of Alexander Library, EGI, Oxford)

C17.   W. S. Green painted a watercolour of Rockall, viewed from the southwest, from which this chromolithograph was made. Green circumnavigated the inaccessible rocky islet in the North Atlantic with an Irish expedition in 1896. (Courtesy of Bodleian Library, Oxford: (per 1996 d549 Plate IX)).

C18.   The kea is a parrot found only in New Zealand. It lives high in the mountains and collects its food by scavenging. W. S. Green found keas to be very inquisitive when he visited Mount Cook. (Courtesy of Alexander Library, EGI, Oxford)

C19.   Jaguars were common in the Andean rain forests in the last century. Howard Saunders, who crossed the Peruvian Andes on foot and by mule in the early 1860s, was fascinated by jaguars. One encounter he had with a swimming jaguar ended up with the jaguar in the canoe and everyone else in the river.

C20.   Green became familiar with grizzly bears while surveying and climbing in the Selkirk Rockies. These bears were drawn by J. J. Audubon who is better known for his illustrations of birds.

C21.   Case of Waterton's birds (Courtesy of Wakefield Art Gallery and Museum)

FOREWORD

'You seem to have taken on a mammoth task', the surviving relative of one of the naturalists featured here wrote to me, 'my impression is that the nineteenth century was composed largely of travelling naturalists!' Indeed, one of the hardest parts of preparing this book was to select a small sample of the numerous naturalists spanning a period of a hundred years. The next problem was to condense their full and varied lives into chapters of manageable length.

Although they lived a century or more ago, these men and women are readily of interest to the modern reader. The obstacles and hardships they had to overcome during their travels have many parallels in today's journeys, whilst the enthusiasm and devotion they showed for their work is shared by many modern field biologists. Since the turn of the century, the focus of field work has changed; we no longer need to stock whole museums with valuable collections of skins and specimens. In addition, field work has become less the domain of the inspired amateur and increasingly that of the highly qualified research worker. For this reason it is easy to view some of the techniques used by the naturalists in the last century with disapproval. In this age of laws to protect wildlife from international trade, for example, it is hard to condone collection on the scale it was carried out by Howard Saunders or John Speke (see chapters 6 and 7.). However, the naturalists' behaviour should not be condemned without being considered in the context of the period in which they lived. From any perspective, these were a unique group of people who helped natural history to emerge into the scientific study of biology.

I received help in many different ways with the preparation of this book, and I am particularly grateful to my mother Ulrica Lloyd for drawing the maps. Alan Eager of the Royal Dublin Society, Timothy Collins of University College, Galway, and Gordon Watson of Wakefield Art Gallery and Museums aided my research; and Lars Lutton of Athens, Ohio, prepared the majority of the illustrations. I should also like to thank the library staffs of Trinity College, Dublin, of the Edward Grey Institute of Field Ornithology and the Bodleian Library, Oxford, and of Ohio University, Athens, for their help.

CHAPTER ONE
The Golden Age of the Naturalist

1.1 Tribesmen from near Lake Tanganyika, from Richard Burton's account of his travels

We had just sat down to breakfast, I was in the act of apologising for appearing barefoot and in a check shirt, alleging, by way of excuse, that we were now in the forest, when a negro came running up from the swamp and informed us that a large snake had just seized a tame Muskovy duck. My lance, which was an old bayonet on the end of a long stick, being luckily in the corner of the room, I laid hold of it in passing, and immediately ran down to the morass. . . . A number of trees had been felled in the swamp, and the snake had retreated among them. I walked on their boles, and stepped from branch to branch, every now and then getting an imperfect sight of the snake. Sometimes I headed him, as he rose and sank, and lurked in the muddy water . . . At last, having observed a favourable opportunity, I made a thrust at him with the lance; but I did it in a bungling manner, for I only gave him a slight wound. I had no sooner done this, than he instantly sprang at my left buttock, seized the Russian sheeting trousers with his teeth, and coiled his tail around my right arm. Thus accoutred, I made my way out of the swamp, while the serpent kept hold of my arm and trousers with the tenacity of a bull-dog.

Thus wrote Charles Waterton, an early-nineteenth-century visitor to Guyana. He was among the first in a succession of men and women who, during that century, chose to abandon the comforts and constraints of English society in order to travel. His special purpose in making long journeys through

South America was to be able to see and collect for himself the animals and plants of a whole new continent. He was a naturalist; one of those who set the study of natural history upon a course which was to lead it to zoology and botany as we now know them.

In order to appreciate the progress of the study of natural history in Britain up to the beginning of the nineteenth century, one must turn to the previous decades. The only branch of natural history to flourish during the eighteenth century was botany, which was founded and fostered by the apothecaries. Popular interest in science was non-existent. Even the best naturalists were disheartened by the high cost of books and other printed information. The difficulty of travel, which often involved tedious journeys by coach or on horseback, prevented them meeting together regularly to compare specimens or exchange ideas. Natural history was a hobby of the wealthy who could afford subscriptions to, or indeed frequently patronised the publication of, the often expensively illustrated contemporary books. At the same time, natural history was being encouraged by the clergy as a suitable pastime for their congregations, and it already appealed to eighteenth-century ladies, many of whom became avid insect collectors. Otherwise it held little or no popular appeal and, even within academic circles, the study of natural history was aimless. The colonisation of foreign lands and the growth of overseas trade brought a steady flow of exotic specimens into museums and private collections. But many of these, lacking a logical order in which they could be arranged, soon became unmanageable hoards.

All this changed at the end of the century with the arrival, accompanied by considerable controversy and consequent publicity, of the system of classification and nomenclature proposed by the botanist Carl Linnaeus. Linnaeus was born in Sweden and worked for most of his life in Uppsala, until his death in 1778. Before his time, the only way of classifying, or arranging, plants and animals had been hopelessly complicated. Every species was given an often unintelligible, long Latin name which attempted to describe the diagnostic features separating it from other species.

Under the new Linnean system of binomial nomenclature, each species had just two names for which the internationally understood Latin was still used. The diagnostic features were provided in a precise and detailed description of the original or type specimen which was published for every species. The nomenclature and corresponding hierarchical classification reflected evolutionary pathways and showed how different species were related to one another. It proved to be a simple but effective system, and is still in use today. Species (and the more rarely used subspecies) represent groups of organisms which are the most closely related to each other. Species are grouped into genera so that two species from the same genus are more closely related than two species from different genera. The hierarchy continues up through families, orders, classes and phyla.

As the Linnean system gradually became accepted at the end of the eighteenth century, natural history studies acquired an important new purpose. Collections were re-sorted and specimens renamed. Soon the demand for new material to fill in the gaps in the hierarchy was growing. A new incentive to finding species previously unknown, and therefore unnamed, was provided by the fact that the discoverer of a species was required to publish a description of the type specimen and to give the new species a name. Under the Linnean system, a species name ending in *-i* or *-ae*, for example, denotes the discoverer's name or the person after whom the species was christened, whilst *-iensis* usually shows where the species was first found.

With these developments came a rash of new specialist societies catering for the more dedicated naturalists for whom no academic training was yet available. The earliest of these was the Linnean Society which published its first *Transactions* in 1791. The Zoological Society of London was founded in 1826, and this gave rise to a separate Entomological

1.2 *The Grant's gazelle was discovered by John Hanning Speke in 1860, and named after his travelling companion*

Society in 1833. The British Association for the Advancement of Science, the Botanical Society of London and the Edinburgh Society were also formed at about this time. By the middle of the century, local clubs and even more specialised societies, such as the British Ornithologists' Union founded in 1858, were thriving.

This expansion in the organised search for knowledge took place alongside an explosion of popular interest in natural history. In many ways, it filled the role of an ideal pastime, as seen by members of early-nineteenth-century society. They were still reeling under the influences of the Evangelical Revival started by Archbishop Wilberforce late in the previous century, and, if it was to be accepted and approved of, an occupation had to be seen to be morally uplifting and healthy, and to involve a certain amount of spiritual exercise. The appeal of health giving excursions into the garden or countryside to examine the wonders of nature created by God was just what was needed. A dedication to natural history was seen to lead 'through Nature to Nature's God' as the theologian William Paley later expressed it. Fortunately, studying nature was also good fun, and it also appealed to the inherent human desire for collecting things.

Another reason for the public becoming infatuated with so scientific a hobby was that there were, in fact, very few professional scientists in the field; almost no one could belittle their efforts at identifying, collecting or writing. Also, just at this time, books became cheap, and more and more people were learning to read them. Mainly responsible for these changes were the development of the steam driven printing presses between 1810 and 1814, and the repeal of the tax on paper and the stamp duty on newspapers, both of which had been imposed during the Napoleonic Wars. In the 1830s, printed publications became accessible to a majority of the population for the first time ever.

The school reforms of the 1850s and the consequent heightening of literacy standards ensured an ever-increasing readership for the popular press. And natural history was a popular subject. Old books, such as Gilbert White's *The Natural History of Selbourne* first published in 1788, were rediscovered and read avidly. There was a stream of new books, natural history journals and manuals, and as demand grew, many became bestsellers. Philip Gosse's *The Romance of Natural History* went into at least 13 editions and spanned 30 years, while his most successful book, *The Aquarium*, made him an unprecedented £900. His *Handbook to the Marine Aquarium* appeared in 1855 and sold 2000 copies within a few weeks. When travellers returned from abroad with tales of exotic plants, animals and peoples, their books also became popular successes. The contemporary taste for the exotic in art and literature was catered for by these extraordinary accounts of the unknown.

Part of the secret of these books' popularity was also their illustrations, some of which were in beautifully lifelike colours. Early books were illustrated with woodcuts. Those by the late fifteenth century artist Albrecht Dürer are possibly some of the best known of the early nature illustrations. At about this time metal plate engraving was also being perfected, and this provided another suitable technique for book illustrations. Copper plate engraving began to be used in the early seventeenth century and from around 1810 onwards this technique was gradually replaced by the comparatively inexpensive lithography or stone engraving. Colour illustrations were obtained by tediously hand colouring the prints, although later the more lifelike chromolithographs were used instead. These were made by superimposing a series of monochrome engravings of the picture, each using a different colour.

The improvements in transport, including Macadam's tar surfacing and the expansion of the railways, ensured that travel grew steadily cheaper and easier, and even tolerably comfortable. The seaside and countryside became accessible to those who lived in the rapidly growing urban areas. Naturalists no longer had to work in relative isolation from each other's encouragement and ideas. With the improvement in communications came an increase in competitiveness amongst collectors, and a sharpness to the criticism of new theories.

The crazes for collecting followed one another in swift succession, orchestrated by popular writers such as Gosse and James Sowerby, and by the desire to put names to the specimens or perhaps even to discover a new species. The seaside and its marine life, which could be dredged, scooped and poked from rock pools at low tide without having to do anything more unsuitable than lift up one's skirts or wet the shoes, provided the focus of attention for most of the 1820s. The need for something stronger than a hand-held magnifying glass with which to examine the plants and animals collected on the shore or elsewhere triggered a demand for microscopes.

Previously these had been rare and unattainably expensive, but 'evenings at the microscope' became the height of 1830s fashion.

The removal of taxes on glass opened the way for the invention of the Wardian display case. Stuffed birds and other animals, dried flowers or growing ferns were collected and arranged in appropriately artistic, but unfortunately not always very lifelike, poses in these glass cases without which no 1830s drawing room was complete. Soon the case was being replaced by another new invention, the aquarium. In these the marine life so carefully accumulated on visits to the seaside could be shown off. By the 1850s large public aquaria had opened and proved to be most successful commercial ventures for their owners. These led in the 1880s to the establishment of the first marine biology laboratories at Plymouth, St Andrew's and Liverpool. But by then, dispirited by the competition from the public aquaria, the public's attention had faded or turned away from natural history altogether.

The obsession with collecting and identifying plants and animals affected the nineteenth-century travellers just as much as those who remained at home. The Grand Tour of Europe had been popular with the fashionable and wealthy in the previous century. As trade with distant countries expanded in the early 1800s, travel became less exclusive and increasingly involved merchants, businessmen and diplomats. There were also voyages of exploration to discover how countries and their produce could be exploited; and surveying expeditions to distant coasts to map new areas or correct errors in existing charts. Members of such expeditions inevitably included some naturalists, and they continued to collect as busily abroad as they had done at home. Others who made these journeys grew fascinated by natural history simply because of the exotic animals and plants they found surrounding them.

Among these travellers were some of the more dedicated and experienced collectors. Sometimes they were actually appointed to expeditions as the official naturalists, or else they maintained an interest in natural history whilst carrying out their normal work. There were a handful of them who were keen enough to undertake the expense, discomfort and dangers of foreign travel solely in order to experience new wildlife at first hand.

The travelling naturalists, with their strong, almost fanatical, common interest, were unusual not only because they voluntarily endured the hardships of long journeys. Many of their colleagues, the so-called closet naturalists, never ventured into the field at all. They were content instead to discover their new species by sifting through the specimens already in museum collections, or by trying to describe the specimens they received from the field naturalists, though they had never seen the animals alive or in natural surroundings. One familiar example of the kind of errors to which this work often led concerns the birds of paradise from New Guinea. As Lynn Barber puts it in her book *The Heyday of Natural History:*

> The famous case of the legless Birds of Paradise . . . arose because all the first Birds of Paradise sent to England had their legs cut off to facilitate packing. Leglessness thereupon became enshrined as a characteristic of the species, and the popular writers went into rhapsodies at the thought of the little creatures spending all their lives in the air. Only the eventual arrival of a Bird of Paradise complete with legs put paid to these ethereal fantasies.

It is easy to imagine how the attitudes and opinions of many of the closet and field naturalists often became sharply and competitively divided.

In fact, prior to incarceration in their respective closets, most indoor naturalists had been active in the field. One way in which a naturalist could travel abroad and collect his own specimens was to join an overseas expedition. Charles Darwin, for example, was taken aboard HMS *Beagle* in 1831 as an 'unfinished naturalist' and gentleman' companion for the captain, Robert Fitzroy. The *Beagle* was commissioned by the Admiralty to survey and chart the coasts of South America, and to make a series of chronological measurements around the world so that longitude could be calculated more accurately. The account of Darwin's adventures and his gradual development into a geologist and biologist are told in his book *The Voyage of the 'Beagle'* published in 1845.

Another naturalist, Thomas Huxley, joined HMS *Rattlesnake* as an assistant surgeon on a surveying expedition to the coasts of Australia between 1846 and 1852. Fortunately for the crew, the demand for his services as surgeon did not fully occupy his time, and Huxley soon interested himself in the contents of the trawls being brought aboard the ship. He became a self-taught marine biologist. he was also an accomplished artist, and his sketches were used to illustrate the account of the expedition. This was published in 1852 by another naturalist called John Macgillivray, the son of a Scottish artist William Macgillivray who had collaborated with J.J. Audubon on his illustrations of British birds made 20 years earlier. A contemporary of Huxley's, Joseph Hooker, whose father Sir William Hooker had established the botanical gardens at Kew in south London, was another travelling naturalist. He accompanied Ross's expedition to the Antarctic in 1839, before assuming his father's duties which occupied the rest of his life.

John Hanning Speke (see chapter 6) accompanied Richard Burton on expeditions of exploration in 1854 to East Africa, and in 1856 when they reached Lake Victoria. He returned to Africa again in 1860 with James Grant, determined to discover the elusive source of the White Nile. Despite his concern with geographical exploration during these journeys, Speke found time to collect wildlife specimens and to make notes and sketches of the animals he saw or shot in the areas through which he travelled. He was exceptionally fond of shooting game but he was also an observant naturalist; this combination of interests turns up in several of the nineteenth-century travellers, and can even be seen in some modern biologists.

Sir Leopold McClintock was one of the people who travelled as part of their work but still managed to make valuable collections and observations of wildlife (see chapter 4). An Irishman by birth and an Admiral in the British Royal Navy, McClintock made important collections of plants, animals, rocks and fossils whilst exploring the Arctic. He was the leader of the expedition which, in 1859, finally succeeded in proving what had happened to Sir John Franklin and his men who had disappeared some years earlier. McClintock's interest in wildlife continued throughout his life. Later, when he was Commander-in-Chief of the British colonies in the West Indies, he made large collections of animals from Barbados and Jamaica.

The British army provided a good means of travel for naturalists, as Speke had found out. In India, Captain Beddome, an army officer, and F. Day, a medical officer, were both able to supply the British Museum with specimens during the 1860s. The diplomatic service had similar advantages. Albert Günther at the British Museum received specimens from consuls as far apart as China, where Robert Swinhoe collected fish and birds for him, and East Africa where the Consul of Khartoum, John Pretherick, was another of his correspondents. In 1867 Günther co-authored a book, *The Fish of Zanzibar*, with Robert Playfair who was consul in Zanzibar and also interested in local wildlife.

Yet another opportunity for travel was sometimes provided by the Church, whose demands on its clergy were often far from rigorous. The Rev. H. B. Tristram travelled widely in North Africa and Palestine in the 1860s and collected birds for the British Museum, whilst Rev. R. T. Lowe provided specimens from Madeira and Portugal at about the same time. Even Rev. W. S. Green (see chapter 9) took time off from his parish of Carrigaline in southwest Ireland to make long climbing and collecting expeditions to the Alps, New Zealand and the Canadian Rockies. This intriguing character and compulsive mountaineer was also among the pioneers of deep-sea marine biology.

The majority of the travelling naturalists

chose to travel just so that they could observe and collect animals. One of these was Mary Kingsley (see chapter 10), an enthusiastic amateur zoologist and anthropologist. In 1893 and 1896, she made long journeys, completely alone, through some of the most remote and hazardous areas of tropical West Africa — quite an achievement for a Victorian lady. Howard Saunders (see chapter 7) spent six years in South America collecting and observing birds. Later he also travelled extensively in Europe, and became a distinguished academic ornithologist, writer and expert on taxonomy. Among his contemporaries was Henry Seebohm (see chapter 8), another ardent collector and geographer who remained a professional businessman all his life. He visited Siberia in 1875 and 1877, travelling as far east as the Yenesei River and inland to Omsk, while collecting and watching birds and other animals. His specimens, including 16,000 bird skins, were given to the British Museum after his death.

Perhaps one of the best known naturalists of the period is Charles Waterton (see chapter 3)

1.4 The range of intermediate forms between two particular species of butterfly of the genus Heliconius *found by Henry Bates in the Amazonian rain forests*

1.3 The Yurak tribe, of which this hunter is a member, lived along the Yenesei river in eastern Siberia

whom we met briefly at the beginning of this chapter. His eccentric escapades and habits, not least his propensity for climbing trees, has endeared him to generations of his readers. Waterton determinedly rejected Linnaeus' system of classifying and naming animals. Instead he gave colloquial names to the many interesting specimens he collected and preserved during his journeys through northern South America from 1812 to 1824. Henry Bates also travelled through South America, but 25 years after Waterton (see chapter 5). He spent 11 years from 1848 onwards collecting on the River Amazon, travelling from its mouth to its uppermost reaches and along many of its tributaries. The enormous collection of insects, birds and mammals he sent back to England was sold to provide him with funds for these journeys.

These are a few of the travelling naturalists whose contribution of specimens and first-hand knowledge helped to make natural history into a science during the nineteenth century. There are many others, too numerous to mention in detail. They include Alfred Wallace, Bates's companion in South America who afterwards travelled to Indonesia and Malaysia; Paul du Chaillu who alarmed and enchanted Victorian society by producing the first descriptions and specimens of gorillas, collected in equatorial Africa; Philip Gosse, already mentioned as a popular writer and illustrator, who travelled in Newfoundland and Jamaica; Henry Johnston, the discoverer of the okapi, a rare forest-dwelling giraffe from Africa; Père Armand David who went to China in 1865 and found a species of deer which was subsequently named after him but has never since been seen in the wild; John James Audubon, the famous artist and book illustrator, who travelled all over North America and Europe; David Douglas, a botanist and pioneer traveller in the American far West, and many others. They are united by the unique period in which they lived that gave rise to their particular brand of amateur scientist and dauntless explorer. The very nature of their travels inevitably led to all sorts of adventures, sometime dangerous, sometimes amusing, but reflecting their own characters: brave, extremely tough and determined, often eccentric and, above all, unprecedented and unparalleled among travellers ever since.

1.5 Paul du Chaillu wrote several extremely popular books on animals, in which they are always described as fierce, bloodthirsty or menacing (Courtesy of the Bodleian Library, Oxford)

CHAPTER TWO
On Being a Travelling Naturalist

2.1 Explorers in East Africa: Burton and Speke

Pick up almost any magazine today and you will see headlines such as 'In the land that produced the wallaby, wombat, emu and kangaroo, you'd hardly expect an ordinary vacation' or 'Discover the natural wonders of Kenya' topping splendid pictures of azure seas crashing onto weird cliff formations or giraffes against the sunset. The advertisements succeed because they conjure up within what, for most of us, is limited experience, those strange animals, landscapes and people. They catch our imagination even although nowadays we all have some idea of what the other side of the world looks like. Bombarded as we are with alluring advertising, television wildlife spectaculars, lavish film sets and superbly illustrated travel books, it seems that nothing will ever again be truly exotic.

In the early nineteenth century travellers had none of this detailed knowledge about their destinations. If they were lucky, an earlier visitor to the same place had published a diary or other account of what he had seen. These varied in style from being lively and readable, to dry and intolerably dull or wandering and anecdotal.

There were, of course, no colourful and interesting photographs in the text. The best the reader could hope for were a few woodcuts or lithographs made either from the author's own sketches or interpreted by one of the currently popular book illustrators, such as J.M.Whymper. A few special editions contained hand-tinted lithographs or even chromolithographs, but the cost of reproduction tended to keep book illustrations to a minimum, sometimes limiting them to just a portrait of the author in the frontispiece. With the increasing availability of cameras during the last decades of the nineteenth century, some new books and fresh editions of earlier ones began to include some photographs, in monochrome and often sepia tinted. Some of these were originals taken by the author, as in Mary Kingsley's books for example, or photographs of painted portraits. Before printing techniques enabled the reproduction of actual photographs in books, portraits were often engravings taken from photographs.

Such books and articles in journals or magazines must have been read and reread by each traveller as he considered his journey, much as we might scrutinise brochures when planning a holiday abroad. The paucity of information on foreign countries added to the air of unpredictability and adventure about the trip which fascinated the would-be traveller, but probably horrified most of the family and friends he prepared to leave behind.

The incentive for travel, then as now, was a highly personal one. Ever since the medieval pilgrimages and crusades of 900 years ago (and even before then), men and women have constantly made long and dangerous journeys simply to satisfy their urge to travel. The desire to escape from the suffocating nineteenth-century society, the need to see scenes described by previous visitors, the longing to explore a country perhaps with the chance of answering the controversial questions of its topography which puzzled those at home, one or more of these reasons must have featured strongly in the minds of the travelling naturalists. At the same time, each had his or her own personal reasons for wanting to travel. The incentive which unites them was their urge to find and collect specimens of plants, insects, birds and other animals, and to discover for themselves the magnificence and intricacies of strange floras and faunas.

Once a destination had been decided upon, preparations for the trip were relatively simple by today's standards. Travel was undertaken alone or in small groups of people, and with relatively little luggage compared with the large manpower and variety of gear needed to mount a modern expedition. One reason for travelling light was the lack of available equipment, although travelling light is a misleading description since the equipment that did accompany a naturalist into the field was often heavy and bulky. William Green and his two fellow climbers, for example, carried loads totalling 130lb with them to the top of New Zealand's highest mountain range. Their tent with waterproof sheeting and poles weighed 26lb, three sleeping bags and an opossum rug another 25lb, clothing and waterproof coats 33lb, and the precious camera and glass photographic plates 33lb. How much easier life would have been with gear, like rucksacks, tents, clothing and food containers, made from the modern durable and lightweight artificial materials, and how much more comfortable if effectively water-proof or insulating fabrics had been available.

The journeys also would have been considerably safer if effective medicines could have been used prophylactically, or to treat diseases and accidental injury. As it was, Mary Kingsley describes how her friends, full of advice and good wishes, showered her with

> various preparations of quinine and other so-called medical comforts, mustard leaves, a patent filter, a hot-water bottle, and last but not least a large square bottle purporting to be malt and cod liver oil, which, rebelling against an African temperature, arose in its wrath, ejected its cork, and proclaimed itself an efficient but not too savory glue.

Clothing presented another problem. Late Georgian and Victorian fashions certainly did not lend themselves to flexibility for the wide range of temperatures and humidities encountered. At the same time the travellers feared any alteration in their familiar appearance or behaviour in case it should suggest that they were conforming to the native standards of dress and decorum. Howard Saunders offers the following advice on the subject for would-be visitors to the Andes:

> For clothing, merino, thin flannel, or silk

next to the skin; a thicker set for the Sierra; whilst for good all-round wear there is nothing like navy serge. In civilised society a black frock coat, waistcoat, and trowsers is as full-dress as any traveller is likely to want... For walking, shoes of *mantas* (hide); the Sierras, strong boots and waterproof overalls, not forgetting the universal poncho, and the necessary blanket, to be stowed under the saddle... And above all take a plentiful supply of that essential commodity which occupies so little space — patience.

In addition, female travellers had to choose between resorting to the more practical but unsuitable male attire, or retaining their dignity and their petticoats. The charming Isabella Bird, for example, who travelled extensively in the Colorado Rocky Mountains in the 1870s (and later in many other remote

2.2 *Isabella Bird in her 'Hawaiian riding dress'*

parts of the world), was horrified and infuriated by a suggestion in a review of her book in *The Times* that she had 'donned masculine habiliments for her greater convenience' while riding on horseback. In order to retrieve her reputation, she insisted a sketch of herself suitably clad 'in Hawaiian riding dress' be included in later editions of the book. Mary Kingsley, faced with the problem of what to wear, admitted to donning a pair of her brother's trousers for her journeys through the wilds of equatorial West Africa. But 'with ineradicable Victorian modesty' she also wore a skirt over the top of them. On one occasion, when Mary fell fifteen feet into a game-trap lined with twelve-inch ebony spikes, she noted with satisfaction: 'It is at times like these that you realise the blessing of a good skirt.'

Achieving the expedition's primary purpose of observing and collecting wildlife was often hampered by lack of equipment. Essential pieces of luggage were the guns and other collecting tools, dissecting instruments, and all that was needed for preserving and storing specimens. Thermometers, barometers and compasses were used but prismatic binoculars, not patented until 1859, were expensive and difficult to obtain. Few travellers could be bothered to carry bulky telescopes with them although these were used at sea. Sharp eyesight and keen hearing had to be relied upon for finding and observing animals. Shot guns were used, rather than live traps, to collect most of the animals, and local natives brought in a variety of specimens which had been trapped or shot with bows and spears. The beginning of the nineteenth century had seen the use of the first detonators in guns, replacing the flint and spark; these were patented in 1807. About the middle of the century, breech-loading guns and, later still, self-contained cartridges were introduced to England from France. These inventions gradually replaced the old muzzle-loaders and greatly improved the collector's chance of success.

Insects could be caught by the aid of a muslin butterfly net and sugared traps, but lighted moth traps do not seem to have been popular. Killing-bottles came into use for insects about 1845 and soon the newly discovered chloroform was being employed in them. From 1854 onwards the use of potassium cyanide became more common. This method of killing insects quickly and cleanly was a welcome alternative to crushing them or pinning them alive. Small specimens of all kinds and the minute details of larger ones could be examined with a magnifying glass, but most of the naturalists also managed to include a microscope in their baggage. Although the microscope had been invented in the sixteenth century, until the 1830s it was far too rare and valuable an item to risk on a journey. The sudden craze for microscopes described in chapter 1 precipitated a plunge in price and, from the 1850s onwards, they were readily available at a cost of between 30 shillings and four guineas.

Small specimens, such as insects and spiders, could be preserved simply, mounted on pins and arranged in pest-proof boxes for storage. Birds and mammals usually were dissected, skinned and cured with arsenic salts. Specimens of intermediate size, and soft-bodied or aquatic species had to be preserved intact. Fortunately the excise duties imposed on glass in England during the Napoleonic Wars at the beginning of the nineteenth century were removed in 1845. This led to an abundance of cheap storage jars and bottles, made of glass and with cork bungs. The specimens were placed in these and some kind of spirit added to stop decay. Local spirit was often used for this purpose, and it was not uncommon for specimens from the West Indies and South America to arrive in England preserved in rum, or those from Scotland and Ireland in whisky!

Other equipment depended largely upon the traveller's own interests. Charles Darwin, for example, bought a barometer and a geological compass or clinometer to take with him on the *Beagle* in 1831. The latter he used to measure the dip and strike of rock beds in order to record respectively their inclination from the horizontal and the direction in which they ran. He also took with him his microscope, of which he was particularly proud because it measured only six inches high by four inches deep. Mary Kingsley was supplied with a collecting outfit by the British Museum on her second trip to West Africa in 1896, partly as a tribute to the discoveries she made on her first visit. Modern discoveries also made the traveller's life more comfortable. Although deliberately travelling as light as he could, Seebohm was accompanied through Siberia in the late 1870s by a pair of 'Cording's india-rubber boots' which he found 'invaluable'.

Each collector had to solve the considerable problem of what to do with his specimens. In most cases, this decision had to be made before leaving home since the crates of storage jars and boxes were to be dispatched to England at every opportunity. Sometimes the specimens never made it home. One of the cases which Speke sent off arrived in London in 1861 with the bottles smashed, the spirit evaporated and the specimens spoilt. Transportation of specimens was always a risky business. Alfred Wallace was shipwrecked in mid Atlantic whilst returning from four years' collecting in South America, and lost his entire collection of live animals and specimens. Joseph Hooker also lost his collection in a fire on board the ship bringing him home from Iceland.

If a caretaker could be found to receive and check the cases as they reached England, so much the better. Darwin was refused storage space successively at the British Museum, the Zoological Society and the Admiralty for his specimens from the *Beagle* voyage, but finally he was able to arrange for them to be sent to his old friend John Henslow, professor of botany at Cambridge. Faced with the same problem, Speke dispatched his specimens to the Royal Geographic Society and the care of P.L. Sclater, and Mary Kingsley entrusted hers to A. Günther in the British Museum.

An essential part of the traveller's work was to record in a journal or diary details of the

2.3 An alligator attacks Henry Bates's camp on a mid-river bank in the Amazon

journey, the animals seen and the specimens collected. Sketches were used for interesting finds and surroundings. Speke, for example, even found time to make watercolours of the East African landscape. Photography was rarely used by travellers in the nineteenth century, although Mary Kingsley was one of those who was able to illustrate the published account of her journeys with photographs.

In the late 1850s, the daguerreotype process had been replaced by the wet colloidal process which was both more light-sensitive and provided a negative allowing for multiple prints from one photograph. The new process was quicker and produced better results than daguerreotype, but it was extremely dangerous due to the flammable nature of the chemicals involved, and the fact that the plates had to be dried over a low flame. The chemicals were also heavy and bulky to transport, and the traveller needed a portable darkroom for on-the-spot developing. Timothy O'Sullivan was a geologist and photographer who surveyed and photographed the Rocky Mountains and Sierra Nevada of North America between 1867 and 1874. He solved the problem by mounting his darkroom and chemicals on a wagon. Although clumsy and slow, this enabled him to take some of the earliest photographs of the far West mountain ranges.

By the last decades of the century, the dry plate process had replaced the wet-colloidal process for photography. Although the equipment was still bulky, the invention was welcomed. 'Photography has become such a help to the traveller,' wrote Green in 1881, '[I] provided myself with . . . 4 dark slides to hold 8 gelatine plates four and a half by three and a quarter inches, 150 plates, and chemicals for developing.'

Nearly all the travelling naturalists were basically amateurs, at least when they started out. These were before the days of the professional field biologist backed by an academic training to guide his collecting. The vast fund of information and array of species the travellers obtained on their journeys often were responsible for making their reputations and future careers. This was certainly the case with Darwin, Thomas Huxley, Bates, Seebohm and Saunders, rather in the same way as it would be true today of a relatively unknown wildlife cameraman, for example, who returns to a television company with an exceptional film.

An important part of their preparations was to obtain as much professional advice as possible, usually from the closet naturalists at home. Charles Waterton was offered and, surprisingly enough, took advice on travel in the tropics from Sir Joseph Banks who had sailed around the world with Captain Cook. Darwin turned to Henslow for advice on preparing flowers and insects, and for the dimensions of the preserving bottles he should order. He went to Adam Sedgwick, professor of geology at Cambridge, for a list of books to take with him on the five-year journey. Sir Francis Beaufort at the Admiralty and the *Beagle's* captain, Robert Fitzroy, told Darwin what equipment he should have and his friend William Yarrell, a naturalist and author who ran a bookstore in London's Piccadilly, helped him buy the items, and also gave instructions on the care and preparation of birds and fish. Bates and Wallace, in their turn, approached Henry Doubleday in the British Museum for entomological advice before they departed for South America in 1848. They also paid a visit to the Chatsworth Hall horticultural collection 'to gain information about orchids, which they proposed to collect in the moist tropical forests and send home.' Kingsley was encouraged by Dr Guillemard, a friend at Cambridge, to collect wildlife specimens on her travels, and together they decided that fish would be most suitable for her.

In addition to gathering up equipment and advice for an expedition, a traveller had to find financial support. There were no scholarships, grants, or travel bursaries on which they could rely, no funding by commercial companies in return for advertising, and little or no government backing. Instead, they often had to invest their own money in the trip and, for most, this represented a sizeable portion of their savings.

Some of the costs could be covered by the sale of the specimens collected, and by the earnings made on published accounts of the journey, but the amount of these modest returns rarely offset the cost of the expedition.

Bates, for example, depended upon funds from the sale of his specimens during his years in South America, and received just 4d. for each insect. One consignment of 7553 insect specimens earned him £125. 17s. 8d., providing a net profit of only £26. 19s. 0d. for 20 months' work. Mary Kingsley chose to travel as a trader in West Africa, exchanging cotton materials and beads for ivory, palm oil and rubber which she later sold. This was partly in order to provide herself with a reason for travelling which the local people could understand, and partly because she needed the money it brought. Others were financially more fortunate. Whilst travelling with Burton in 1856, Speke remained in the pay of the British Army from which he had taken three years' leave due to him. Leopold McClintock too was paid officially, by the Royal Navy, during his Arctic voyages.

Finally, the logistic problems facing travellers in the nineteenth century were not unlike those experienced today. Although there were no passenger liners, overseas trade routes were sufficiently well established to ensure that ships left Britain regularly for most major foreign ports. Passage could be procured on these if the traveller was prepared to forgo most of the comforts which are regarded as the minimum necessary on a passenger vessel. Once ashore, guides and porters were hired from among the native population, and provisions purchased locally. Most of the naturalists made use of friends and letters of introduction in order to obtain accommodation for themselves and their men along the route; and, more important still, to collect advice on the whereabouts of the best hunting grounds for specimens in the immediate vicinity.

If things went wrong, however, there was little that could be done to summon help. In fact there was often no way the travellers could even be found amidst the forests, mountains, tundra or savannah through which they passed. Sir John Franklin and his men and ships, for example, could not have disappeared so completely in the 1980s in the Canadian Arctic as they did in the 1840s. A re-enactment of Franklin's voyage was undertaken in 1967 by a Canadian Armed Forces expedition with full logistic backing including helicopter support. They covered the whole area searched over by McClintock's party (see chapter 4), but found no more traces of Franklin.

2.4 This veil was used to avoid mosquitoes on the Siberian tundra

Places visited by Charles Waterton in Guiana (modern Guyana) and other parts of South America, 1804–24

CHAPTER THREE
The Wanderer of Walton Hall

3.1 Walton Hall, home of the Waterton family for over four centuries, until Charles' son, Edmund, was forced to sell the house and estate in 1876

One morning in 1789, a boy of seven climbed onto the roof of a sturdy Georgian house. If caught, he was in trouble and he knew it, but the lure of a wheezing nestful of young starlings had been too much for him. He was the eldest of the six children in his family, and rather prone to showing off. He had once swallowed a skylark's egg whole in order to impress his sister, and had been punished with an emetic of mustard administered by his mother. On this occasion, however, he was more fortunate. He was discovered on the roof not by his angry mother but by the cook who lured him down with a piece of gingerbread.

This child was the young Charles Waterton, destined to be one of the earliest and most eccentric naturalists of the nineteenth century. He had started life just as he intended to continue it, and as Dr Hobson, Waterton's special friend and his family's physician, tells:

'When Mr Waterton was 77 years of age, I was witness to his scratching the back part of his head with the big toe of his right foot.' Charles Waterton's escapades throughout his varied life have delighted his readers and biographers for well over a hundred years, and only a small sample of them can be included here. He had an enthusiastic interest in natural history which lasted all his life. For example, we hear of him dodging prefects at school in order to go birds'-nesting and, even at the age of 82, climbing the trees of a heronry on his Yorkshire estate to count the herons' nests.

His journeys took him to South America, the West Indies, North America and Europe. He was always searching for specimens and was willing to add birds, mammals, reptiles and in fact any kind of vertebrate or invertebrate to his collection. Many years after his death, Waterton's main biographer, Philip Gosse, wrote:

It may at once be stated, without fear of contradiction, that Charles Waterton was not a man of science nor had he a scientific mind. Although a keen and close observer and describer of nature, and an excellent field naturalist, his greatest admirers could not describe Waterton as a scientist. In fact, he himself always held what he disdainfully dubbed the 'closet naturalist' in the greatest contempt.

Charles Waterton was born on 3rd June 1782, the fourteenth direct descendant to occupy the family mansion of Walton Hall, near Wakefield. The Watertons were an ancient English Catholic family who traced their ancestors back to the Norman Conquest. As Waterton later expressed:

Up to the reign of Henry VIII, things had gone swimmingly for the Watertons; and it does not appear that any of them had ever been in disgrace. But during the sway of that ferocious brute there was a sad reverse of fortune.

At ten, Waterton was sent to a Roman Catholic boarding school in the village of Tudhoe, near Durham. Four years later, he moved to Stonyhurst College in Lancashire, a Jesuit school which had been founded in Belgium in 1592. Here Waterton quickly gained a reputation for climbing, not always limited to trees, as Gosse relates:

Charles hankered after a jackdaw's nest high up in the clock tower. To reach the nest he proceeded to climb up the perpendicular face of the college building ... the Rector, who on looking up and seeing the perilous position of young Waterton, ordered him immediately to come down ... so the reluctant boy clambered slowly down to the ground. The same night, so legend recounts, a loud crash was heard, and the very stone parapet by which Charles Waterton had been about to pull himself up fell head-long to the ground.

Stonyhurst was to be the limit of Waterton's formal education since English universities were barred to Catholics. At Tudhoe, his persistent interest in natural history had been firmly discouraged as being unsuitable; and at Stonyhurst his education had been purely a classical one since the teaching of science subjects simply did not exist. All the same the Jesuit Fathers were remarkably tolerant of his pursuit of birds' nests, and Waterton enjoyed his time at Stonyhurst. For the rest of his life he liked to wear a blue swallow-tailed coat with gilt buttons similar to the school uniform.

A visit in 1802 to two of his uncles who had emigrated to Spain and lived in Malaga gave Waterton a taste of travel — and of unconventional experiences. He survived the notorious earthquake and plague of Malaga in which thousands of people, including one of his uncles, died of yellow fever. Waterton and his brother were given false papers and smuggled out of Malaga on board a Swedish ship to avoid the quaratine regulations. Upon his return, Waterton spent almost a year at Walton Hall fox-hunting, studying the estate's wildlife and helping his father, but soon he was ready to be on the move again. He decided upon a visit to another uncle who owned estates in British Guiana (now Guyana), South America.

In the autumn of 1804, Waterton went to London where he stayed with yet another of his uncles, Sir John Bedingfield. He was taken to meet Sir Joseph Banks, the famous naturalist and dilettante (who had agreed to join Captain Cook's second expedition only if he could bring with him not one but two horn players to provide musical accompaniment during dinner). Now at 61, Banks was the respected President of the Royal Society but still wore his hair powdered. His advice included a tip which Waterton never forgot — he was not to endanger his health by remaining in the tropics for more than three years at a time. Banks remained Waterton's friend and correspondent until his death in 1820, and was just about the only person from whom Waterton readily accepted advice.

Waterton sailed from Portsmouth in

November 1804, and two months later he arrived in Georgetown, on the estuary of the Demerara River. Britain had captured the territory from the Dutch only two years before Waterton's arrival, and the city was a paradise for 'cock-fighting and card-playing gamblers, as well as those votaries of Venus.' Waterton had become a teetotaller on his return from Spain, and because of this he was regarded in Georgetown as very eccentric. He occupied himself in looking after his uncle's plantations of coffee, cotton and sugar, and 'Walton Hall,' a sugar estate on the Essequibo River recently acquired by his father. This gave him plenty of opportunity for studying the wildlife on the nearby estuary and surrounding savannah.

Much of the rest of his time was spent on the estate of his only close friend, a Scottish planter named Charles Edmonstone, who had lived in Guiana for over 20 years. Edmonstone's home was renowned for its feasts of ham and guinea pig, a particular delicacy in South America. Waterton had his own reasons for liking the place. The Mibiri creek close by was a haven for hundreds of scarlet ibises, white egrets, boat-billed herons, pelicans and waders, and the savannah around the plantations was 'a paradise for wildbirds.' It was here, while collecting specimens in the swamps and mud-flats usually avoided by Europeans, that Waterton first caught a fever, probably malaria, which bothered him on and off for the rest of his life.

Edmonstone's wife was the daughter of a Arawak Indian princess and a Scottish colonist. As a result, Edmonstone commanded great respect among the local Indians who helped him find special timber in the forests, and protected his estates from attacks by vengeful escaped slaves. In 1809, Waterton attended the christening of the Edmonstone's second child, Anne Mary, and is said to have declared that one day, with her parents' permission, he would marry her.

The year Waterton arrived in Demerara his father died, and he became the twenty-seventh

3.2 Horned and crested screamers, relatives of the ducks and geese but lacking webbed feet

Squire of Walton Hall. He returned to England several times to arrange the running of the estate in Yorkshire, but he was in Guiana when war broke out between Spain and England in 1807. In September, despite his religion and to his great pride, Waterton was commissioned by the Governor of Demerara as a lieutenant in the second regiment of the militia. He spent some time in Barbados, headquarters of the army and navy and centre of their social life, and thus, Waterton felt, possessing little to interest a naturalist. The following year Waterton was sent with despatches from Demerara to Angostura (now Ciudad Bolivar) on the Orinoco River in central Venezuela. This was his first experience of primary tropical rain forest. He thought it 'a grand feast for the eyes and ears of an ornithologist', an opinion with which it is easy to agree. Thousands of waterbirds flocked in the swamps and on the wooded islands along the river, and the calls of parrots and macaws rang through the canopy. Loudest of all was the trumpet-like voice of the horned screamer, a bird resembling a turkey but actually related to the ducks and geese.

In 1812, now aged 30, Waterton decided to give up his managerial responsibilities on the Demerara plantations and travel through the unknown forests in the interior of Guiana. Inevitably, he hoped to add to his natural history collection, but the main aim of the expedition was to find an Indian tribe which made the famous wourali poison, now known as curare. Waterton had the idea that, if collected in its pure state, the poison might provide a remedy for hydrophobia or for lockjaw, more familiar to us as rabies and tetanus. These are both diseases in which violent pain and ultimately often death is caused by involuntary muscular contractions. Waterton knew wourali was a muscle relaxant and killed by paralysis so, he reasoned, it might be a suitable antidote. Another of his ambitions for the trip was to discover the legendary city of El Dorado, on the lake of Parima which had been described fancifully by Sir Walter Ralegh and other travellers.

The Spanish *conquistadores* of the sixteenth century had explored most of Guiana in their rapacious search for gold, but no maps of anything other than the coastline existed in

3.3 'No animal seems capable of withstanding the united attacks of the Peccary', wrote a contemporary of Waterton's

3.4 Waterton's accounts of animals, such as the sloth, were unlike anything published before because he drew upon his observations of their natural behaviour in the wild

Waterton's time. It was not until Sir Robert Schomburg explored the interior of the country two decades later that the first reliable maps became available.

Waterton left Georgetown in April with six Indians and a negro, and travelled up the Demerara River, soon leaving behind the orderly riverside plantations. The forest was full of jaguars, tapirs, peccaries (wild pigs), red howler monkeys, anteaters, sloths, vampire bats, caimans, snakes and numerous birds including 'scarlet curlews' or scarlet ibises. Waterton's description of the sloth, which with 'his looks, his gestures, and his cries, all entreat you to take pity on him', illustrates how he liked to write about the animals he saw. His accounts were usually acutely observant but, to modern taste, highly anthropomorphic. However, the fact that he gave copious descriptions of animals in their natural surroundings, rather than of their skins or stuffed bodies, was a great novelty for his contemporary readers.

As Waterton travelled up the river, the sounds of the forest drowned the rushing water. From the treetops came the plaintive whistle of the tinamou, the yelping toucan, the shrill kiskadee flycatcher and the resonant tolling of the bell-bird. By night, they heard the different kinds of frogs which 'almost stun the ear with their hoarse and hollow sounding croaking', the owls and goatsuckers which 'lament and mourn all night long' and the howler monkeys 'moaning as though in deep distress'. The party left the Demerara near a series of falls, over 120 miles from the river's mouth and, carrying the canoes, they took a path to the Essequibo River which they found flooded and impassable. Instead, they followed a tributary of the Essequibo to the Macoushi country where the local Indians killed their game with wourali-tipped arrows and spears. To his delight, Waterton succeeded in collecting several small samples of the poison.

As he neared the Brazilian frontier, Waterton wrote a formal letter to the Portuguese governor of the outpost at Fort Saint Joachim, but was denied permission to cross the border and pay him a visit. Almost as soon as the party had set off down the Takutu River, Waterton had a severe attack of fever. This time the governor responded, and Waterton spent a week being nursed and recovering his strength and curiosity in the Portuguese fort. The governor knew nothing of El Dorado or Lake Parima, despite his 40 years in the area, so, bearing his precious samples of wourali, Waterton set off downstream again.

The return journey was made particularly difficult by heavy rain, accompanied by lightning and pounding thunder, and Waterton's recurring fever. The canoes shot the rapids on the Essequibo River where, as Waterton describes, 'the roaring of the water was dreadful; it foamed and dashed over the rocks with a tremendous spray, like breakers on a lee-shore, threatening destruction to whatever approached it.'

At last they arrived at the Edmonstones' house where Waterton remained to recuperate. He learnt that the sudden loud rumbling which had alarmed his camp on the night of the first of May had been a volcanic eruption on the island of Saint Vincent, over 500 miles away in the Lesser Antilles.

Waterton returned home to England in 1813, via Grenada and Saint Thomas's in the West Indies, but it was to be three more years before he finally shook off recurrent bouts of fever. Meanwhile he was forced to turn down a tempting invitation to explore the island of Madagascar in the Indian Ocean. He experimented with the wourali he had collected, killing the first donkey into which it was injected in under 12 minutes. Another animal to which the poison was administered was miraculously revived by an incision into its windpipe and the artificial ventilation of the lungs with a pair of bellows which circumvented the paralysed respiratory muscles. The donkey was sent to Walton Hall after the experiment and there, named Wouralia, she lived for another 25 years.

The 260-acre Walton Hall estate is claimed to have been the first British nature reserve. Waterton did everything in his power to make the animals and birds feel at home there, and to protect them from poachers and the hated rats. This included building a ten feet high wall around three miles of the park's boundary. Much to the disapproval of his superstitious housekeeper, he built a ruin, which soon became covered with ivy, for the barn owls to nest in. But Waterton adds: 'I assured the housekeeper that I would take upon myself the whole responsibility of all the sickness, woe and sorrow that the new tenants might bring to the Hall.'

In March 1816, Waterton started out on the second of his 'wanderings', as he called his journeys to South America. He landed at Pernambuco (known today as Recife) near the easternmost tip of Brazil. The city looked charming from the sea but its streets were narrow, dirty and neglected. The forest, a short distance from the town, was full of birds although it was the rainy season and most were moulting. He collected 58 specimens, carefully selecting the best-looking ones and, as a result, ending up with a rather unscientifically biased sample consisting of mostly adult males rather than the inconspicuously coloured females.

The ships which called at Pernambuco were crowded because, as Waterton puts it, 'transporting the poor negroes from port to port for sale pays well in Brazil'. Eventually he was able to embark in a Portuguese brig and two weeks later landed at Cayenne, on the coast of French Guiana. He went first to visit the clove-tree plantations and botanical gardens at La Gabrielle. Here he met the director, Monsieur Martin, who had travelled through much of the Far East collecting plants for the gardens, and who had sent over 20,000 botanical and zoological specimens to Europe. Determined to add a tropicbird to his collection before he left Cayenne, Waterton set out in a canoe for a desolate offshore stack where both tropicbirds and frigatebirds nested. First the canoes became lost in the maze of creeks leading to the coast, next they almost filled up with water in a torrential rainstorm, then they were grounded by the falling tide in the middle of huge mud flats and the men had to wait all day long under the tropical sun for the sea to return. But at high tide the sea became too rough for the canoes to reach the stack and the expedition had to be called off.

Other plans also had to be shelved at this

stage. Waterton had hoped to travel south again, and up the Amazon from Pará (Belém), then on to the Rio Negro, and thence overland to the source of the Essequibo River where he intended to renew his search for El Dorado. Unfortunately, the coastal current sweeping north past Cayenne was so strong that, in Waterton's words, 'a Portuguese sloop which had been beating up towards Pará for four weeks, was then only half way.'

Instead he took an American ship north from Cayenne to Paramaribo on the coast of Dutch Guiana (now Surinam), and travelled on to New Amsterdam and Demerara. Here he returned once again to the rain forest with the following advice to would-be fellow travellers:

> Leave behind you your high-seasoned dishes, your wines, and your delicacies; carry nothing but what is necessary for your own comfort... A sheet, about twelve feet long, ten wide, painted and with loop-holes on each side, will be of great service: in a few minutes you can suspend it betwixt two trees in the shape of a roof. Under this, in your hammock, you may defy the pelting shower, and sleep heedless of the dews of night. A hat, a shirt, and a light pair of trowsers, will be all the raiment you require. Custom will soon teach you to tread lightly and barefoot on the little inequalities of the ground, and show you how to pass on, unwounded, amid the mantling briars.

During this part of the trip Waterton paid special attention to the many different kinds of birds, and his account of the journey included characteristically detailed description of the appearance and habits of hummingbirds, cotingas, the bell-bird, toucans, toucanets, motmots, trogons, cassiques, jacamars, troupials, manakins and many other birds of the Guiana forests. Waterton was well aware that he was seeing these most beautiful of the world's birds in all their true colours and not in the way his contemporaries knew them, as shrunken and faded museum specimens. The toucans, for example, lost the colours of their huge and spendid bills when skinned and stuffed. 'About eight years ago, while eating a boiled Toucan,' Waterton recalls, 'the thought struck me that the colours of the bill of a preserved specimen might be kept as bright as those in life.' A little careful dissection with his penknife showed him that if all the inner layers of the bill were removed, leaving only the outer horny sheath, discoloration could be prevented.

Waterton spent six months in the forest and collected over 200 species of birds. Best of all, his health had been excellent all the time. In April 1817, he set out across the Atlantic for England. En route he spotted one of the prized tropicbirds sitting on the sea within range of his gun. He shot it without much thought of how it was to be retrieved, but saying hopefully to the crew 'A guinea for him who will fetch the bird to me.' Almost instantly a Danish sailor plunged fully clothed from the forecastle. Both he and the bird were narrowly saved from 'Davy's locker', and Waterton had his specimen of a tropicbird.

Soon after his return to Walton Hall, Waterton began to consider another journey, this time with an expedition to Africa to explore the river Congo. He went to London and met the 'scientific men' on the expedition, advising them on 'the absolute necessity of temperance; and... never to sleep in their wet clothes.' One cannot help wondering how the liaison would have worked out for Waterton was by habit a solitary traveller. Anyway, Sir Joseph Banks took Waterton aside and warned him against joining the expedition which he said, quite rightly, would not be successful. Instead, Waterton returned to work on the estate, making holes in a ruined wall for breeding starlings, planting copious ivy and holly bushes, clipping yews to encourage small birds to nest, and making an artificial sand martin colony in a quarry. From a telescope in the drawing-room of the house he could count the waterfowl on the lake.

That autumn he departed for a holiday in Europe with a friend, named Alexander, who was a captain in the Royal Navy. In Rome they met another old friend of Waterton's from his

Tudhoe school days who, it appears, could match Waterton's climbing skills. For fun the two men scaled the cross on the very top of Saint Peter's and left Waterton's gloves on the tip of its lightning conductor. This was not the last they heard of it:

> News of these strange proceedings on the part of two eccentric Englishmen quickly reached the ears of the Pope, Pius VII, who ordered the gloves to be removed instantly, since they rendered the conductor useless. But a difficulty arose, for nobody could be found bold or agile enough to attempt the task, so that, in the presence of a vast and delighted Roman audience, the Squire had to go up again and fetch down his gloves.

Late in 1819, Waterton was called to Sir Joseph Banks's deathbed. Typically, their final conversation concerned the shortcomings in the current methods of taxidermy. The lips and noses of mammal specimens, they decided, should be removed and replaced by wax models.

Waterton was full of ideas of how to improve the preservation and mounting of his specimens, and was most critical of museum exhibits: 'Were you to pay as much attention to birds, as the sculptor does to the human frame, you would immediately see, on entering a museum, that the specimens are not well done.'

He was particularly disappointed by the fact that many of the museum buildings being built in newly industrialisd towns of England and other European countries were being filled with beautiful picture galleries and ornamental libraries, while the contents of the natural history museums was being 'left to pure chance':

> Museums ought to be encouraged by every means possible. The buildings themselves are, in general, an ornament to the towns in which they have been built; whilst the zoological specimens which they contain, although prepared upon the wrong principles, are, nevertheless, of great interest; since they afford to thousands, who have not the means of leaving the country, a frequent opportunity of seeing rare and valuable productions which are found in far distant parts of the globe.

The main problem with museum exhibits lay in the methods of preservation of specimens currently in use. Animals and birds collected in hot or humid countries were skinned quickly and often carelessly, and then dried as fast as possible before they could be attacked by rot or marauding insects. This process shrank and distorted the skins which had to be forced into some kind of realistic pose by stretching over a wire or iron armature and stuffed. Specimens had to be thrown away when poor preparation of large animals or even insects and eggshells led to stinking exhibits.

One of Waterton's first discoveries was that turpentine put inside stored skins or soaked in a sponge and hung in a specimen chest effectively deterred insects. He was able to prepare and mount his specimens without using stuffing. He placed great emphasis on delicate and painstaking skinning, but his recommended techniques were considered by many of his contemporaries to be too time-consuming. Insects, for example, could be preserved complete with their vivid lifelike colours if they were dissected segment by segment, the body contents removed, and the pieces sewn back together again. It was essential, Waterton found, to be equally meticulous when skinning larger specimens. Every bone in the body was removed down to the tips of the toes and tail, and even the wattle or legs of the tallest bird were carefully skinned. Extra thicknesses of skin such as that in the animal's lips, nose and soles of its feet were pared away. The fur, feathers or scales were washed before skinning began, and care was taken to keep them clean and to avoid stretching the skin.

The next stage, as Waterton put it,

> now depends upon the skill and anatomical knowledge of the operator (perhaps I ought

to call him artist in this stage of the business), to do such complete justice to the skin before him, that, when a visitor shall gaze upon it afterwards, he will exclaim, 'That animal is alive!'

The skin was stuffed lightly with sawdust, and the correct body contours were moulded using needles and gimlets pushed through small holes in the skin. A thread was used to replace the support of the backbone. Even the lips, ears and face were remodelled until the specimen was lifelike in appearance and had been arranged in a realistic pose, which only someone who had seen the animal in the wild could know about. It was then allowed to dry slowly, without contortion. The sawdust was drained through slits in the skin of the soles, and glass eyes were inserted by a slit under the jaw. Waterton was particularly proud that his technique enabled the specimen to stand up on its own and did not require either stuffing or wire armatures.

The real secret to Waterton's method lay in not allowing the skin to dry out during preparation whilst at the same time preventing putrefaction. The accepted technique was to cure the skin with 'arsenetical soap', the use of which sometimes had lamentable consequences, as Waterton relates:

> I knew a naturalist, by name Howe, in Cayenne, in French Guiana, who lost sixteen of his teeth. He kept them in a box, and showed them to me. On opening the lid — 'These fine teeth,' said he, 'once belonged to my jaws: they all dropped out by my making use of the *savon arsenetique* for preserving the skins of animals'.

Instead Waterton discovered he could cure specimens either before or after skinning by immersion in a solution of 'a good large teaspoon of corrosive sublimate' which he mixed in a bottle of wine. Corrosive sublimate or mercuric bicholoride is a strong and acrid poison, actually little safer to use than arsenic. When dissolved in alcohol, Waterton found, it penetrated the tissues of any specimen, both preventing decay and killing insects. With typical ingenuity Waterton also used his corrosive sublimate solution to preserve or waterproof his top hats, coats, trousers, the interior of his carriage, and the furs and ostrich feathers on the bonnets of his sisters-in-law.

In February 1820, Waterton set out on the third journey of his 'wanderings.' He travelled back to Demerara, and returned to the Edmonstones' house on Mibiri creek. The family had left for Scotland three years earlier, and the abandoned house was half ruined. Waterton had the roof restored and settled in to continue his studies of natural history. He spent the rest of the year collecting in the forest, the overgrown plantations and the swamps by the creek, and in perfecting his taxidermy techniques on even the most demanding of specimens. These included a large caiman, the capture of which gave Waterton more notoriety at home than any of his other exploits. He seized an opportunity to jump astride the caiman and grab its forelegs. Avoiding its thrashing tail, he pulled it onto a sand bank. There, he tells us, 'I cut his throat; and after breakfast was over, commenced the dissection.'

At Sir Joseph Banks's suggestion, Waterton was preparing three lectures on taxidermy, one on insects and reptiles, one on birds and one on quadrupeds. But on his arrival at Liverpool from Guiana in April 1821, his ten precious boxes of specimens were impounded by the Customs who demanded that their contents should be revalued. Waterton refused and set out in disgust for Walton Hall with nothing but a pair of live Mallay fowl given to him in Georgetown which he was to breed on his estate. After six weeks of argument, Waterton gave up and paid the extra duty, and his collection was delivered safely to Walton Hall.

For three years Waterton stayed peacefully on his estate, concerning himself with the welfare of its birds and animals. Then, in 1824, he received a copy of *Ornithology of the United States* written by Alexander Wilson, a Scottish weaver who had emigrated to America and

3.5. This portrait of Waterton's caiman ride was painted by Captain E. Jones, an early school friend (Courtesy of the Trustees of Stonyhurst College)

died there in 1813. The book and its marvellous illustrations fired Waterton's restless imagination, and soon he was heading off across the Atlantic once more.

In July, he took a steamer from New York up the Hudson River to Albany. It did not take Waterton long to realise that he had come to the wrong place 'to look for bugs, bears, brutes and buffaloes' as he puts it, but instead he made the best of enjoying the beautiful scenery and its inhabitants, especially of a certain young lady from Albany. To the amazement and alarm of his fellow-travellers, he insisted on curing a sprained ankle he had received when alighting from the coach by holding it in the rushing waters of the Niagara Falls.

Waterton travelled on from Buffalo via Lake Erie to Montreal and Quebec. On the journey he met a solitary bug which he found crawling over his neck. Compared with those in Guiana, it was 'a little half-grown, ill-conditioned' one so, instead of killing it, he threw it into some luggage lying close by and 'recommended it to get ashore by the first opportunity'. Waterton is remembered today in Canada by the Waterton Lakes National Park in southern Alberta which extends into Montana. Although Waterton never travelled this far west, the lakes were named after him in 1858 by Thomas Blakiston, an English Royal Artillery officer and naturalist who was a member of the Palliser Expedition searching for a route through the Rocky Mountains.

After returning to Albany and thence to New York, Waterton set out again to visit Philadelphia, the city in which Wilson's *Ornithology* had been published. Here he visited the Museum of Natural History, one of the first of its kind in the country, which had been set up by the artist Charles Wilson Peale. One of the collection's finest exhibits was a mammoth found by Peale which was, in Waterton's view, 'the most magnificent skeleton in the world'.

Heading for Guiana once more, Waterton left the United States for Antigua, Guadeloupe and other islands in the West Indies. On Dominica he admired the enormous edible frogs, the six-inch rhinoceros beetles, and a particularly fabulous purple hummingbird. Waterton was unable to settle down to his study of wildlife until he had regained his beloved Demerara forests and embarked on the fourth of his 'wanderings'. He turned again to describing the three-toed sloth of which he was so fond. The animal had never been observed in the wild before, and it was Waterton who pointed out that specimens should always be drawn and mounted in a tree rather than on the ground.

'Of all the specimens Waterton brought back with him from the forest,' wrote his biographer of the fourth visit to Guiana, 'none was to create such a sensation and so much speculation as the "Nondescript", and nothing Waterton ever did caused more harm to his reputation as a serious naturalist.' Waterton claims it to have been an enormous animal with a thick coat of hair and great length of tail. He was able to bring home only the head and shoulders of the beast. In fact, the Nondescript was both a practical joke and testimony to his skills as a taxidermist. He had taken the head and shoulders of a red howler monkey and modelled the face to resemble human features. The skin turned black when it dried and contrasted horribly with the wild ginger hair. Waterton put the specimen on show in Georgetown and convinced many of the hundreds of people who went to see it that it was a so-called jumbi or wild man of the woods. Later he proudly displayed a picture of his Nondescript as the frontispiece of his book where it attracted all kinds of attention, criticism and delight. The pamphleteer, Sydney Smith, recognised it immediately as a political pun, and claimed to have noticed the head 'basking in the House of Commons after he has delivered his message'.

In December 1824, the rainy season began

3.6 *Waterton developed a new method of taxidermy which he applied to this specimen, the Nonedescript, prepared as a practical joke (Courtesy of Wakefield Art Gallery and Museum)*

and Waterton headed for England with his specimens, so ending his last journey to South America. On his return to Walton Hall, he settled down to prepare an account of his travels for publication. The style of his diaries suggests they were written with the public in mind. Closely observed accounts of the animals with which he became familiar during his life on the Demerara plantations are interspersed with lively tales of his trips into the interior. To these he added an essay on taxonomy which presented the views he would have delivered in lectures in 1820 had his specimens not been delayed by the Customs. The book appeared in 1825 under the lengthy title *Wanderings in South America, the North-West of the United States, and the Antilles, in the years 1812, 1816, 1820, and 1824. With original instructions for the perfect preservation of birds, etc., for cabinets of natural history.* (Although Waterton never went any further west in the United States than upstate New York, this was nearly the limit of the populated northwest at the time.) The book proved so popular that a second edition was published two years later and subsequent editions have continued to appear ever since, the last one in 1973.

Whilst many of the book's critics ridiculed its style, its content also stirred up controversy. As Gosse explains: 'There were two subjects in particular which stuck in the throats of the scientific critics: the story of the riding on the back of a cayman and, worst of all, the Nondescript. The Nondescript they never forgot.' The American ornithologist William Swainson was particularly vociferous in his disbelief of Waterton's caiman ride, and this started a lifelong feud between the two men. Despite the book's shortcomings, it was not possible to disguise what Sydney Smith called 'the genuine zeal and inexhaustible delight with which all the barbarous countries ... are described.'

At four in the morning of 11th May 1829 Charles Waterton was married to Anne Mary Edmonstone whom he had held in his arms at her christening 17 years before. Both Charles Edmonstone and his wife had died shortly after returning from Guiana in 1817, and their three daughters had been sent to school at Bruges in Belgium. It seems likely that Waterton had acted as guardian to the children although his marriage to Anne, literally straight from school, comes as a surprise to even his biographers. Within a year of the wedding, Mrs Waterton died, just three weeks after giving birth to a son. Waterton was overwhelmed with grief; he became very religious and rather a recluse. He moved to a small room at the top of the house where he studied, and prepared his specimens. He also slept there, lying on the floor wrapped in an Italian military cloak with a block of wood for a pillow. Anne's two sisters, Elizabeth and Helen, moved into Walton Hall to care for the baby, and soon a happy household was established.

Some years later Waterton's taste for climbing spread to scaling rock faces as well as trees. He visited the great sea cliffs at Bridlington on the Yorkshire coast where he collected eggs of guillemots and razorbills from their precarious nesting ledges on the sheer rock. In 1840, Waterton set out on a Grand Tour of Europe accompanied by the Edmonstones and his small son Edmund. The journey, orientated by the cities which possessed museums of natural history, took them through Holland, Belgium and France, to Rome and Sicily. At each museum Waterton took delight in criticising the taxidermy of the exhibits.

Waterton spent the last 20 years of his life at Walton Hall where he became affectionately known as The Squire. He continued to improve the grounds for wild birds and animals, and even designed new kennels and stables to make life better for the domestic animals. He also kept up a constant attack upon what can best be described as his pet aversions. His numerous letters to the naturalist George Ord of Philadelphia, for example, were filled with criticism of those he regarded as closet naturalists such as Swainson, J.J. Audubon and his colleague William MacGillivray.

When Waterton's *Essays on Natural History* was published in 1871 it included a lengthy biography of the author by Norman Moore, later President of the Royal College of Physicians. Moore describes how he met Waterton first in 1863, when as a 16-year-old medical student he had walked all the way from Manchester to arrive uninvited at Walton Hall hoping to see Waterton's museum. He found a severe-looking Georgian mansion on an island at the end of a large lake surrounded by thick woodland and meadows. The house was reached by a narrow iron bridge and there he met the 82-year-old Waterton, a tall, very thin old man with close-cropped white hair.

Moore was left alone to examine Waterton's magnificent art collection, acquired on impulse in Würzburg, West Germany, and including pictures by Michelangelo, Van Dyck, Holbein, Rembrandt and Rubens. The top landing of the house, its staircase and hall were filled with cases of Waterton's beautifully prepared specimens, birds from the markets in Rome, from the estate and from Guiana, various crustaceans and numerous stuffed animals. Overhead, hanging from the ceiling, was the twelve-feet long caiman which had attracted so much notoriety. Among the specimens Moore found a few hybrids of Waterton's own making including one labelled succinctly 'England's Reformation, Zoologically Illustrated or Old Mr Bull in Trouble'.

This was the first of many visits Moore paid to the Squire. Sometimes Waterton would practise his own brand of practical joke on Moore (and other more unsuspecting visitors to Walton Hall). Waterton would hide under a table in the hall and lunge out to bite the newcomers' legs as they passed by. Together Waterton and Moore made the round of the park checking nest boxes and hollow trunks or climbing trees and walls to examine nests. The friendship lasted until Waterton's death, at which Moore was present. On 26th May 1865, Moore set out with Waterton to fell some alder trees. While crossing a narrow plank bridge carrying a log, Waterton fell heavily. Later that night as he became weaker he asked for the window of his room to be opened so he could hear the corncrakes in the meadows across the moat. At two the following morning, Waterton died surrounded by his grandchildren, sisters-in-law and friends.

His funeral was a splendid, if unconventional, affair. Moore describes the procession:

> Foremost upon the lake went a boat, which carried the Bishop of Beverley and fourteen priests, who chanted the Office for the Dead as they rowed along. Next came a boat which bore the coffin. The boats with the mourners followed, and the procession was closed with a boat, which ... was empty, and draped in black.

Moore's laborious praise of Waterton's character leaves one with the impression of a man fearing neither danger nor criticism, yet compassionate and shy, with enduring vitality and great good nature. In Moore's opinion, Waterton had never obtained his rightful place as a man of science. His writing was what could now be regarded as one of the earliest attempts to popularise biology. Waterton's avoidance of Latin names, for example, which so frustrated his fellow scientists trying to use his texts, was due to his fear that 'a repulsive nomenclature would scare away the public'. Waterton was not only 'an unwearied outdoor observer, as well as a diligent dissector'; unlike many of his contemporaries, he was aware of the importance of comparative anatomy, and of the adaptation of structure to function in animals.

Perhaps Waterton's most important contribution, and one for which he is now considered 'the father of modern scientific and artistic taxidermy', was to develop new methods of treating specimens which not only ensured their long-term preservation but also made museum exhibits lifelike in appearance and pose. One bird specimen he prepared in Demerara stayed in the tropics for two years and was then taken to England where it remained for five months. Then it was returned to Demerara but taken back again to England where after five more years, Waterton claimed, it was still unfaded and unchanged.

The Walton Hall museum was dismantled after Waterton's death. Almost all his specimens, including one new species of hummingbird, *Thalurania watertonii*, collected in Brazil, and his collection of Indian artifacts, were cared for by his old school, Stonyhurst College. They can now be seen, on permanent loan from Stonyhurst, at Wakefield Museum in Yorkshire. There are around 750 items in Waterton's collection, and the old 'Squire of Walton Hall' no doubt would be pleased to know that numerous mounted animals and birds are still in good condition.

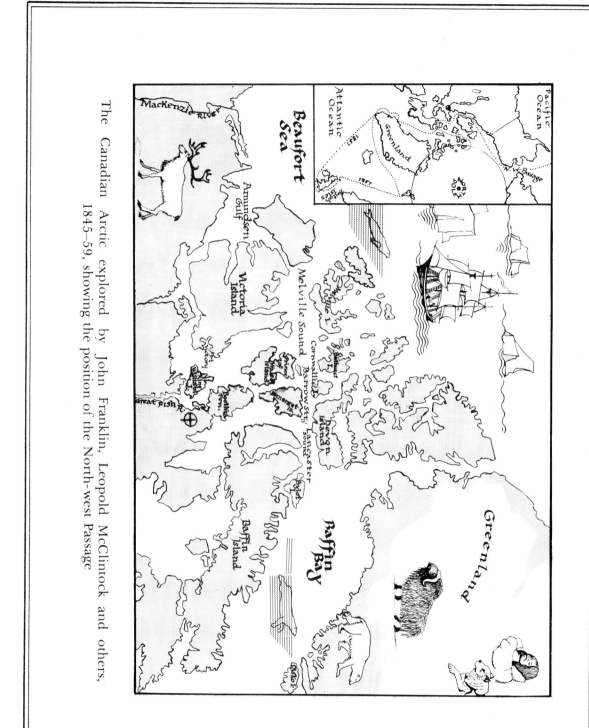

The Canadian Arctic explored by John Franklin, Leopold McClintock and others, 1845–59, showing the position of the North-west Passage

CHAPTER FOUR
The Fate of Sir John Franklin

4.1 Sir Roderick Impey Murchison gave support and advice to many explorers and collectors, including John Franklin (Courtesy of the National Portrait Gallery, London)

In December 1844, Sir John Barrow, a veteran Arctic explorer, approached the First Lord of the Admiralty with ambitious plans for a new expedition. In Elizabethan times, an Englishman, William Baffin, had discovered the gateway to the Northwest Passage through the Arctic. Now it was time, Barrow argued, for Englishmen finally to complete the search for a way through the treacherous Canadian pack-ice to the Pacific. In doing so, England could not only claim the honour of the discovery, but also be the first to reap the rewards that this new trading route would bring. On the other hand, the feat 'if left to be performed by some other power,' Barrow assured the Admiralty, would mean that England 'having opened the eastern and western doors, would be laughed at by all the world for having hesitated to cross the threshold.'

The idea was well received, and Barrow's proposals were scrutinised by Arctic authorities including the Royal Geographic Society, Sir Edward Parry, Sir James Clarke Ross, and Sir John Franklin. The latter had made his reputation for Arctic travel in 1819 by his overland exploration of the coast of the Beaufort Sea west of the Mackenzie River. In his history of polar exploration, Laurence Kirwan wrote:

The adventures and gallantry of these men of the Royal Navy in such an unusual and romantic setting deeply stirred the British people. Their daring descents in frail canoes over foaming rapids and swirling whirlpools . . . all these moving episodes contributed to a national legend about the heroism of John Franklin.

There seemed to be no one more obviously suited than Franklin to lead the Government's new expedition. Two of the Navy's ships, HMS *Erebus* and HMS *Terror*, were refitted and strengthened for pack-ice, and an enormous quantity of stores and equipment was assembled. In May 1845, the expedition set sail from Greenwich, in east London. No one doubted that Franklin and his crews were embarking on a successful voyage to discover the Northwest Passage, and there was a general air of optimism and excitement. As Sir Roderick Murchison, President of the Royal Geographic Society, affirmed, the expedition would do everything 'for the promotion of science and for the honour of the British name and Navy that human efforts can accomplish'. As it turned out, Murchison was entirely correct, but not in a way anyone could have expected at the time.

At the end of July, the captain of a whaler in Lancaster Sound off Canada saw the two ships moored to an iceberg, and he spoke to Franklin. After this the expedition simply disappeared. No-one, it seemed, had seen or heard anything more of the two ships or their crews of 129 men.

The details of what had befallen the Franklin expedition became known gradually during the next twenty years. Throughout this time authorities on polar exploration, and many others who knew nothing of it, became intrigued and even obsessed by the mystery. Franklin had sailed from where he was last seen in Lancaster Sound into Barrow Strait. Then he turned north, possibly on the recommendation made to him by Parry before he had left England. The expedition passed round the north of Cornwallis Island and, finding no way through to the west, moved south again and back into Barrow Strait.

The *Erebus* and the *Terror* were laid up for the winter of 1845–6 at Beechey Island off the southwest tip of the huge Devon Island. When spring arrived, they headed south between Prince of Wales and Somerset Islands but soon entered heavy pack-ice. By 12th September 1846, both ships had become inextricably wedged in the sea ice flowing south from the broad expanse of Melville Sound. The expedition had travelled nearly half what we now know to be the Northwest Passage, and had almost reached the Boothia Peninsula. In fact, had they been able to travel south and then westwards along the mainland coastline, they would have reached the open waters of Amundsen Gulf and the Beaufort Sea.

On 11th June the following summer, Sir John Franklin died and Captain Francis Crozier took command of the expedition. A second winter was spent trapped in the groaning grinding pack-ice and, in April 1848, they were forced to abandon the ships which were breaking up under pressure from the ice. Only 105 officers and men had survived, and most were weakened by scurvy. They set out over the frozen sea towards the coast of King William Island, hoping to follow it south to the Great Fish River (also called the Black River), and to be able to reach an outlying trading post of the Hudson's Bay Company. They pushed on valiantly but left a trail of abandoned equipment, graves, and corpses in their wake. Not a single man from the expedition had survived.

In the years that followed, successive expeditions set out from both England and America to look for Franklin and his men. Although their voyages provided some of the missing pieces in the geographical jigsaw of the Canadian Arctic, they came no closer to discovering what had become of Franklin. In particular, they searched to the north of Barrow Strait and Lancaster Sound, assuming that Franklin would have taken Parry's advice and looked for a passage to the north. Finally, in January 1854, seven and a half years after their disappearance, the Navy proposed that the officers and men of the Franklin expedition should be considered to have died on Her Majesty's Service.

Lady Jane Franklin, Sir John's second wife, was a tenacious and indomitable character. Her portrait shows a beautiful woman in an impossibly low cut dress, with huge blue eyes and a firm set mouth. As the Franklin mystery caught the public's attention and imagination, she became a heroine for her long suffering and courage. A popular ballad, *Lady Franklin's Lament*, was even sold on the streets of London:

My Franklin dear long has been gone
To explore the northern seas,
I wonder if my faithful John,
Is still battling with the breeze;
Or if e'er he will return again,
To these fond arms once more
To heal the wounds of dearest Jane,
Whose heart is grieved full sore.

However, once he and his men were pronounced dead, Franklin was gradually forgotten.

Then in October 1854, some interesting news arrived from Dr John Rae, a surgeon with the Hudson's Bay Company. The Eskimos had told him that a group of 40 or 50 white men had been seen in the spring of 1850 hauling boats towards the Great Fish River. Rae had found remains of 30 men from Crozier's party at the river's mouth. He had been able positively to identify them by the guns and silverware marked with initials and crests that were found with them. The British Government was struggling with the Crimean War and had no time to mount yet another search expedition. Instead, Lady Franklin launched a public appeal for funds, and in 1855, with the money raised, she bought a 177 ton steam yacht, the *Fox*. With such a small ship and crew, the qualities of the captain were obviously critical to the expedition's success. Lady Franklin fortunately made an excellent choice when she asked Captain Leopold McClintock to command her expedition, and the Prince Consort helped arrange McClintock's secondment from the Navy.

Amidst all the responsibilities of his ship and the long voyage, through fair weather and

4.2 Admiral Sir Leopold McClintock

some of the worst the Arctic had to offer, McClintock maintained his fascination for natural history. He had no formal training, but during 11 years of Arctic travel he always found time to record and observe plants, animals, birds, fossils and geological details. He amassed a large collection of rock specimens and fossils, including several species new to science. In addition, he made a valuable contribution to Canadian Arctic geology, and also developed new techniques of sledge travel based largely on what he learnt from the Eskimos.

Leopold McClintock was born on 8th July 1819 in Dundalk, about 50 miles north of Dublin on the east coast of Ireland. He was the second son of Henry McClintock, an Army officer who had become head of Dundalk Customs House, and Elizabeth, a daughter of the Archdeacon of Waterford. He was the eldest of the 12 surviving children born to Elizabeth, 'a very pretty woman, of remarkable ability and energy'. The family's firstborn, Louis, died as a child and 2 younger babies died in infancy. Leopold attended the local school in Dundalk, and led a full and happy life. His father was a popular man, a musician and a good sportsman. The rolling farmland and woods of County Louth and Dundalk Bay, with its vast flocks of wintering wildfowl, must have given young Leopold McClintock a good introduction to wildlife.

All his childhood he wanted to go to sea. His idol, an older cousin named William, was already in the Navy, and as a small child he was profoundly impressed by a print of Admiral Berkley in uniform which hung in his father's dressing room. The third and possibly the most pressing reason for his ambition was that entrance into the Navy did not require a knowledge of Latin!

Before his twelfth birthday in 1831, McClintock's cousin offered him as a naval cadet to Captain Charles Paget. He left home excited and in such a hurry that he took with him only his most valued possessions, with a bag of marbles and a bottle of apple juice. His solitary journey took him to Dublin, onwards by steamer to Bristol, and on a ten-hour coach ride to Portsmouth where, on 22nd June, he presented himself at his cousin's ship, the *Samarang*. Here he was taken in with delight by the crew who pronounced him 'so small that it was like looking for a flea in a blanket'. In fact, McClintock weighed 68lb, just 2lb more that William's Newfoundland dog.

There had been little change in the British Navy since the Napoleonic Wars. The *Samarang* was a small frigate of 113 feet, with a crew of 160 men and only four feet nine inches of space between decks. McClintock took to Navy life immediately, and his cousin, the ship's First Lieutenant, is said to have set a good example to him in matters of drinking and swearing. The work was strenuous, the hours long and not without danger. Within a month of the *Samarang's* departure for South America, McClintock had managed to fall 60 feet from the rigging, fortunately with nothing more than bruises and burnt hands. The food on board was dull but sufficient in quantity and variety to keep the men healthy. For breakfast there was cocoa and ships biscuits. Dinner, at midday, consisted of 'salt junk' or pork with biscuits and, sometimes, pumpkin pie. The last meal of the day, served at five o'clock, was fresh meat, vegetables and fruit tarts if the ship was near port. McClintock soon became a favourite of Captain Paget, and was given sweets from his private supply.

They made landfall at Rio de Janeiro in Brazil after 50 days' sailing, and then proceeded north to Bahia (Salvador) where McClintock was taught to swim in the harbour. In the weeks that followed they rounded the treacherous Cape Horn, and spent three months at Callao in Peru. Here one of the ship's lieutenants, William Smyth, decided to travel over the Andes and down the River Amazon to the Atlantic. The journey had only been done once before by a European, Lieutenant Henry Maw, in 1847. One of McClintock's messmates, Frederick Lowe, was given leave to accompany Smyth. McClintock listened and watched with fascination as the plans and elaborate preparations for the expedition were made. Smyth's account of this journey was to prove compulsive reading for another Englishman, preparing to take roughly the same route 25 years later, Howard Saunders (see chapter 7).

The *Samarang* returned to Portsmouth in January 1835, after three and a half years abroad. The years that followed were happy ones for McClintock. He served on ships in the Channel fleet, in the West Indies and off Newfoundland, and then trained at the Royal Navy College, Portsmouth. Leave in intervals between these postings was spent quail shooting and fishing near Dundalk, or walking in the beautiful Dublin mountains.

A term of duty as mate on the *Gorgon* in 1843 – 5, beginning when McClintock was 24, earned him his first promotion for special service. The steamer became grounded on a sandbank during a storm in the shallow Rio de la Plata estuary at Montevideo. McClintock assessed the situation and suggested successfully how she could be refloated. He earned himself the Captain's highest praises:

He is without exception one of the steadiest, most zealous and excellent young men I ever served with, and is deserving of his promotion or any other favour the Admiralty might confer on him.

In 1847, McClintock applied to join Sir James Clarke Ross's expedition to search for Franklin. Competition for the appointment was stiff, but McClintock was accepted on the recommendation of Captain William Smyth, the Amazon explorer from the *Samarang*. In February 1848, McClintock heard he had been appointed Second Lieutenant to Ross on the *Enterprise*. As McClintock's biographer, Sir Clements Markham, points out: 'This was the turning point in McClintock's life. It was his great opportunity and he seized it.'

Ross had already proved himself an expert in polar exploration. He had been First Lieutenant to his uncle, John Ross, when he commanded an expedition to the Arctic in 1830 with the paddle steamer *Victory*. The men had made long journeys by sledge south to the Boothia Peninsula and King William Island, and had discovered the magnetic North Pole. The *Victory* was stuck for 3 winters in the ice, but the crew finally reached Lancaster Sound where they were picked up by a whaler. James Ross had gone on to make his reputation in the southern hemisphere, in Antarctic exploration. In 1847 he was selected by the Admiralty to lead one of their three expeditions which were to continue the search for the Northwest Passage, and to try and find Franklin.

The *Enterprise* and her sister ship *Investigator* were barque rigged sailing ships. McClintock joined them at Woolwich in February 1848, and met Sir James Ross. He found a 'short, stout, square-built man, with an aquiline nose and very piercing black eyes.' In McClintock's words, 'he seems a very clever, quick, penetrating old bird, very mild in appearance and rather flowery in style.' The two men became great friends, and shared many opinions on the subject of Arctic travel. First Lieutenant on the *Enterprise* was Robert McClure who became another of McClintock's lifelong friends.

The expedition left London in mid May. A month later, off Cape Farewell they saw their first pack-ice and icebergs. It is difficult to explain the immensely forceful impression of ice at sea. The sea, threatening enough with its uncontrollable power over those who live on it or by it, acquires a new but spectacularly beautiful menace. To McClintock 'it was an extremely interesting and beautiful sight'. Soon he was seeing masses of unfamiliar Arctic birds and animals.

They took refuge at Port Leopold, at the northeast corner of Somerset Island, and there they spent the winter of 1848–9. Had they but known it, the expedition was less than 75 miles from Beechey Island where the Franklin expedition had wintered three years before. Port Leopold was, in McClintock's view, 'without exception, the most barren spot with which we are acquainted'. Cliffs of earthy limestone and mudstone stretched for unbroken miles either side of their mooring, and the only birds McClintock shot were two ptarmigan, a few snow buntings and a solitary starving grackle, well outside its normal range. But he found fine specimens of selenite and fibrous gypsum in the shale beds of the nearby cliffs, and discovered many fossils in a bed of 400 million-year-old limestone. These included natural casts of fossilised gastropods (snail-like molluscs) which were later found to belong to a new species, christened *Loxonema mcclintocki* in honour of its finder.

All through the dark uncomfortable winter that followed, McClintock read books on geology and botany, and planned with Ross for the sledge journeys they would make in the spring. Before this time, Arctic exploration had been confined largely to what could be seen from a ship and by short journeys on foot, or slightly longer ones hauling provisions in heavy sledges and even in ship's boats. Ross's previous expedition with his uncle had shown him that long-distance travel by sledge was not impossible. McClintock, with his eye for technique and knowledge of the natural environment, was able to develop the methods they needed to cover vast distances of inhospitable snow and ice.

That spring, Ross, McClintock and their men hauled their sledges 500 miles in 39 days, a record distance. They covered the coast of Somerset Island as far south as Bellot Strait which separated it from the Boothia Peninsula. Two hundred miles of newly discovered coastline had been added to the map, but no sign of Franklin had been found. In reaching Bellot Strait, McClintock had made a valuable discovery. The maps he was using, like those given to Franklin three years before, showed Somerset Island connected to the Boothia Peninsula by a narrow isthmus. Franklin, not suspecting the existence of Bellot Strait, naturally had kept to the west of Somerset Island, rather than sailing into the inlet to the east from which he assumed there was no outlet. Once in the heavy ice flow between Somerset and Prince of Wales Islands his ships had been pushed offshore and away from any chance of noticing the opening of James Ross Strait. This would have led them round the east and south of King William Island in relatively ice-free waters, and through into the remainder of the elusive Northwest Passage.

That autumn both Ross's ships drifted through Lancaster Sound in the pack ice and south into Baffin Bay where the ice broke up releasing them to return to England in November 1849. McClintock and Ross had shown that Arctic exploration on foot and sledge was both feasible and safe, and that men could live off the land using a knowledge of the Arctic fauna, and not have to rely entirely on heavy stores. In August, for example, over 30lb of fresh meat had been provided for the crew of the *Enterprise* by shooting 2000 birds.

McClintock returned to the Arctic for a second time six months later, on an expedition commanded by Captain McClure. This time he was First Lieutenant on the *Assistance*, under Captain Horatio Austin. The ship had been adapted for Arctic travel and this greatly improved the health and comfort of the crew. The *Assistance* reached the edge of the pack-ice in Melville Bay on 1st July 1850, and waited with the whaling fleet for the ice to clear. 'The straining and cracking of the ship's timbers' McClintock later recalled, 'was not a pleasant sound.' At last they were able to proceed into Lancaster Sound where McClintock was amazed by the spectacular scenery of Cape York at the northwest corner of Baffin Island. The cliffs soared 1200 feet vertically from the sea and were covered with screaming seabirds, little auks which McClintock calls 'rotches', 'looms' which were Brunnich's guillemots or murres, and black-legged kittiwakes.

The expedition landed once again at Beechey Island where they discovered the winter quarters and hunting camps of the Franklin expedition. They drifted south down to the tip of Cornwallis Island caught in the sea ice. McClintock, by now an expert ice navigator, hunted bears and continued his collecting, keeping live lemmings on the messroom table. They wintered on the coast of Cornwallis Island, near Griffith Island which McClintock soon found to be rich in limestone fossils. He hunted Arctic foxes, hares and ptarmigan, and later admired the carpets of spring flowers which seemed to emerge from the snow itself, purple saxifrages, yellow Arctic poppies and stunted ground willow. Some of the fossil molluscs collected on Griffith Island belonged to a new species which was later named *Orthoceras griffithi*, after Mr Griffith, 'the founder of Irish geology'. Also among McClintock's finds on Griffith Island was a new trilobite, *Cromus arcticus*, which is an extinct marine arthropod.

Ross's expedition had demonstrated to McClintock the importance of maintaining the crew in good health and spirits through the long winter. Captain Austin trusted McClintock, and allowed him to take some of the crew on sledging trips in the autumn to lay out caches of food and provisions in preparation for an early start the following spring. Throughout the winter months that followed, McClintock insisted that officers and men alike were fed a varied diet, that suitable warm clothing was issued to them, and that exercise was compulsory.

In April 1851, McClintock set out with a sledging team for Melville Island, carrying tents and over 150lb of bear meat. They found plenty of additional food in the form of musk oxen, reindeer, Arctic hares and ptarmigan on the island. The cold was intense. McClintock's diary describes how 'the fat on our salt pork becomes hard and brittle like suet; to drink out of a pannikin without leaving the skin of one's lip attached to it, requires considerable experience and caution'. When the wind was favourable, they were able to speed their progress over the ice by setting a groundsheet on tent poles like a sail.

They camped in Parry's old site at Winter Harbour on the south side of Melville Island. Here, according to McClintock:

We lived on very friendly terms with our neighbour the hare. She regarded us with confidence, hopped about our tent all day, and almost allowed the men — who were most anxious to carry her back to the ship as a pet from Winter Harbour — to touch her.

In the sea nearby, McClintock collected a mussel-like shellfish belonging to a species which geologists knew from fossils found in Scotland and Sweden but had assumed to be extinct. The arrival of summer as they turned back to the ship in June made travel slow. The tundra became covered with flowers and alive with birds as phalaropes, dotterel, sandpipers, brent geese and gulls arrived and began to nest.

When they arrived on board the *Assistance*, after an absence of 80 days, they had travelled over 750 miles. This was quite an accomplishment, even by today's standards. The *Assistance* and the rest of the Admiralty's fleet returned safely to England in October 1851, having covered over 6000 miles of Arctic islands and sea — still without finding Franklin.

McClintock returned to Dublin. There he met the Reverend Samuel Haughton, professor of geology at Trinity College and later known for his ardent opposition to Darwinism, who helped McClintock sort and identify his Arctic collection. The following year McClintock made his third visit to the Arctic, with an expedition of four ships headed by the elderly Sir Edward Belcher. This time McClintock had his own command as captain of the steam tender *Intrepid*. In his crew were several members of his old sledging teams from the *Assistance*, and also a naturalist named Dr Scott.

The *Intrepid*, and the *Resolute* to which she was tender, travelled westwards to Melville Island following the route taken by Edward Parry with the *Hecla* and the *Griper* 33 years before. Caught in the ice off Griffith Island, McClintock collected more plants and fossils, and dredged for marine animals. Nearby, on Bathurst Island he made a valuable collection of fossils from the carboniferous sandstone. The beds of sandstone there and on Melville Island, he found, were quite different from those he had seen either at Griffith Island, Cornwallis Island or Beechey Island. He found different species of fossil coral in each group of sites.

By September 1852, the expedition was preparing to winter in Bridport Inlet on the south coast of Melville Island. McClintock made several sledge trips that autumn and one of his teams found a message from McClure on the *Investigator*, who was facing his third winter stranded in the ice. This discovery led to the rescue of McClure's expedition the following spring.

McClintock and his sledge teams started out from their winter quarters in April 1853, using small fast satellite sledges to explore inlets and to conduct hunting parties. They crossed the centre of Melville Island to the northwest corner. By early May they had reached the large island even further to the northwest, Prince Patrick Island, which they explored thoroughly for the first time. At the point where they had first landed on the island, McClintock found ammonites and many other fossils in a bed of fine clay or lias. One of the ammonites later provided McClintock with yet another new species, named *Ammonites mcclintocki*.

By the time his sledge teams returned to the ships, they had been away over three months and had covered 1400 miles. The winter of 1853–4 was passed in the pack-ice south of Melville Island with the remnant of McClure's crew from the abandoned *Investigator*. As usual, McClintock planned his journeys for the spring, hoping to visit the south side of the strait. He was amazed to receive a message from Belcher, who was trapped in the ice in an exposed position north of Beechey Island, ordering the *Intrepid* and the *Resolute* to be abandoned. McClintock travelled to the ship to argue personally with Belcher against this unnecessary move. He found Belcher 'looking old and debilitated from want of fresh air and exercise' but could not persuade him to change his mind. Even this journey was a feat of speed in sledge travel: McClintock and his companion Thompson took only 12 days to cover 460 miles. With the ships, McClintock was forced to abandon his valuable library and his precious dried plants and fossils.

When the expedition returned to England in October 1854, McClintock learned that, in his absence, his mother had died and war had broken out in the Crimea. However, what interested him almost more was the news from Dr Rae in the Hudson's Bay Company of possible traces of the Franklin expedition. While Lady Franklin's expedition in search of her husband's remains was being organised, McClintock attended to his Arctic specimens. He contributed a chapter on Arctic zoology to the second edition of McClure's account of the *Investigator*'s voyage which was published in 1856. He also gave papers on Arctic travel to the Royal Dublin Society in January and May of 1856. His new sledging techniques had other advantages besides making exploration easier, and he concluded his talk by saying:

> We have seen the gradual expansion from journeys of 500 to 1400 miles . . . There is now no known position, however remote, that a well equipped crew could not effect their escape from by their own unaided efforts . . . By our own experience . . . [we] have secured to all future Arctic explorers a plan by which they may rejoin their fellow-men.

In April 1857, McClintock left Dublin for London to make final arrangements with Lady Franklin. The *Fox* took on board provisions for 28 months including 'preserved vegetables, lemon-juice, and pickles, for daily consumption, and preserved meats for every third day; also as much of Messrs. Allsopp's stoutest ale we could find room for'. As usual, McClintock was determined to keep his men in good health and spirits. The Admiralty prepared over three tons of pemmican for the expedition. This was produced by smoking strips of prime beef over wood, pounding it and mixing in an equal weight of melted beef fat, and pressing the mixture, which would keep for several years, into fifty-pound cases.

Other equipment such as hydrographic instruments, charts, chronometers and ample Arctic clothing were also lent by the Admiralty. The Board of Trade supplied them with 'a variety of meteorological and nautical instruments and journals'. Some of the equipment was the most modern available at the time. 'Mr Clifford, the inventor' presented the expedition with 'an apparatus for lowering a boat at sea' and 'an apparatus for reefing topsails'.

In the *Fox*'s crew of 23 were 15 men with previous Arctic experience including several who had offered to serve when they learned that McClintock would be their captain. The First Lieutenant was William Hobson, whose portrait shows him to have been a young man with a pleasant open face and the kind of bushy beard confined to the underneath of his jaw which was fashionable. The ship's Master was Allen Young from the Merchant Navy. He was closer to McClintock in age than the other officers, and had a low brow and receding hair line with side whiskers and thick drooping moustache. Also in the crew were Carl Petersen, a Norwegian interpreter with all the appearance of a sea captain, and Dr David Walker, the ship's surgeon. The latter was a young man with a moustache joined to his side whiskers and at least six inches long. He was, no doubt to McClintock's delight, an experienced botanist.

McClintock's small party left England at the

beginning of July 1857, rounding Scotland via the stormy Pentland Firth and buffeting through ice south of Spitsbergen. Leaving the southern tip of Greenland behind and heading north up the coast, they ran into thick fog. Then, as McClintock records:

> Suddenly the fog rolled back upon the land, disclosing some islets close to us, then the rugged points of mainland, and at length, lifting altogether, the distant snowy mountain-peaks against a deep blue sky. The evening became bright and delightful; the whole extent of coast was fringed with innumerable islets, backed by lofty mountains, and, being richly tinted by a glorious western sun, formed an unusually glorious sight. Greenland unveiled to our anxious gaze that memorable evening, all the magnificence of her natural beauty.

When they reached Frederickshaab on 20th July, they were able to replenish stocks of fresh food with codfish, a few ptarmigan, hares and a couple of goat kids. McClintock noted in his diary: 'I saw fine specimens of a white swan, and of a bird said to be extremely rare in Greenland — it was a species of grebe, *Podiceps cristatus*, I imagine.' The only swan which reaches west Greenland is the North American whistling swan which breeds in the Canadian Arctic. The great crested grebe (*Podiceps cristatus*) which McClintock mentions is a common breeding bird in England, but not in his native Ireland, and would indeed have been a rare visitor to western Greenland.

A month later, the *Fox* had crossed the Arctic Circle and entered Disko Fjord. McClintock wrote:

> The scenery is charming, lofty hills of trap rock, with unusually rich slopes (for the 70th parallel) descending to the fjord, and strewed with boulders of gneiss and granite. We found the blue campanula holding a conspicuous place amongst the wild flowers. I do not know a more enticing spot in Greenland for a week's shooting, fishing, and yachting than Disco Fiord; hares and ptarmigan may be found along the bases of the hills; ducks are most abundant upon the fiord, and delicious salmon-trout very plentiful in the rivers. Formerly Disco was famed for the large size and abundance of its reindeer; but for some unexplained reason they now confine themselves to the mainland.

Later, as they steamed through a sea like glass, McClintock saw huge flocks of eider ducks which extended for several miles. He noted they were mostly in eclipse or non-breeding plumage; it is likely that they were moulting birds and probably temporarily flightless. Flocks of eiders gather in the sheltered shallow water along the Greenland coast where they can dive for shellfish.

Passing through Melville Bay, McClintock was once again deeply impressed by the 'grandeur of the mighty glaciers, extending unbroken for 40 or 50 miles'. As a geologist, he would have been well aware of the powerful effect of such a mass of ice on the landscape. Towards the end of August, the *Fox* became trapped in the ice of Melville Bay which was exceptionally heavy that year. They drifted slowly southwards, occasionally shooting seals on the ice flows. 'The liver fried with bacon is excellent', McClintock commented. Two

4.3 A Greenlander's supper appropriated by a bear

Eskimos, Christian and Samuel, had joined the crew in Greenland and they now provided valuable hunting skills. Christian and the Norwegian Petersen tried anything to attract seals:

> They scrape the ice, thus making a noise like that produced by a seal in making a hole with its flippers, and then place one end of a pole in the water and put their mouths close to the other end, making noises in imitation of the snorts and grunts of their intended victims.

McClintock admitted that he did not know if this method was particularly successful, but it did look very entertaining!

The weather was becoming extremely cold by September but they were still stuck fast in the ice. McClintock calculated their position was only about 12 miles from open water. They shot 43 seals which they fed to the 29 sledge dogs on board. These had been bought in Greenland as an experiment by McClintock to improve sledge travel. They promptly 'devoured their two days' allowance of seal's flesh (60 or 65lb) in 42 seconds!' At this stage, the dogs were not allowed on the ship although they devoted all their efforts to stealing aboard when no one was looking.

The salted meat, food for the human expedition members, was lowered through the ice to soak for a few days before being cooked. To McClintock's astonishment, one netful was completely devoured by a shark. 'It would be interesting to know how such fare agrees with him, as a full meal of salted provisions will kill an Eskimaux dog which thrives on almost anything,' he wrote.

Late in October, furious gales and a snow storm hit the ship which was quickly housed over. McClintock saw a brilliant meteor passing through Cassiopeia, so bright that it looked like a flash of lightning. As the weeks passed into December, he had plenty of time to think about how the animals survived in Arctic conditions:

> Anything which illustrates the habits of animals in such high latitudes I think is most interesting, their instincts must be quickened in proportion as the difficulty of subsisting increases. Foxes, white and blue, are very numerous; all the birds are merely summer visitors, therefore the hare is the only creature remaining upon which the foxes can prey; but the hares are comparatively scarce: how then do the foxes live for 8 months of the year? Petersen thinks they store up provisions during the summer in various holes and crevices, and thus manage to eke out an existence during the dark winter season; he once saw a fox carry off eggs in his mouth from an eider-duck's nest, one at a time, until the whole were removed; and in winter he has observed a fox scratch a hole down through the very deep snow, to a cache of eggs beneath.

The mountains of Disko Island were sighted again in March as they continued their drift southwards. With the break up of the sea ice in April their situation became dangerous. The jostling, grinding pack threatened to smash the wooden hull of the *Fox*, but at last on 26th April she was free to move out of the ice, first under sail and later steaming. The ship had drifted for 242 days, covering over 1100 miles. Always mindful of his crew's well-being, McClintock ordered a course for Holsteinborg 'reputed to be the best place for reindeer upon the coast.'

They remained two weeks in the pretty, civilised port of Holsteinborg but had to replenish their meat supplies with eiders, seabirds (looms and dovekies), and ptarmigan as no reindeer had been seen that year. The beginning of June found the *Fox* steaming into Melville Bay again. They spent an anxious few hours off Buchan Island aground on a submerged reef which had been hidden by floating ice, but succeeded in refloating the *Fox* with the next high tide. The men landed for the last time in Greenland near Cape York, a point which McClintock described as being of fine-grained granite with quartz and white feldspar. He continues:

4.4 The Fox *aground temporarily off Buchan Island*

Steep slopes of rocky debris, which screen the bases of the most precipitous cliffs, form secure nurseries for the little auk; these localities were literally alive with them; they popped in and out of every crevice, or sat in groups of dozens upon every large rock. I have nowhere seen such countless myriads of birds. The *rotchie*, or little auk, lays its single egg upon the bare rock, far within a crevice beyond the reach of fox, owl, or burgomaster [glaucous] gull. We shot a couple of hundred during our short stay on shore, and, by removing the stones, gathered several dozen of their eggs.

Soon they were heading for Coburg Island on the coast of Canada. McClintock began collecting specimens of some of the pack-ice seabirds, fulmars, three ivory gulls, two Brunnich's guillemots and an Arctic skua. They moved south into Lancaster Sound, past the enormous Cape Hay 'loomery', a vast Brunnich's guillemot colony. Near here, McClintock shot 'an immense bear' over eight feet long. which he says was 'destined for the museum of the Royal Dublin Society'.

On 11th August, the *Fox* and her crew arrived at Beechey Island, a good site for limestone fossils. McClintock erected a marble tablet, provided by Lady Franklin, with an inscription beginning: 'To the memory of Franklin, Crozier, Fitzjames and all their gallant brother officers and faithful companions who have suffered and perished in the cause of science and the service of their country.' They steamed on across Barrow Strait to Port Leopold on Somerset Island, past the spectacular sheer-sided Prince Leopold Island, home for over a third of a million nesting seabirds (and now the site of long-term study by ornithologists from the Canadian Wildlife Service).

The expedition moved south along the coast of Somerset Island to the mouth of Bellot Strait which was already filling up with ice and almost impassable. On 1st September, McClintock took to one of the ship's boats with a small party of men including Dr Walker who, like so many later visitors to the Arctic, came along 'for the novelty of the cruise, bringing his camera to fasten upon any thing picturesque'. They passed safely through the strait, rowing below towering cliffs still whitewashed from the excreta of the thousands of seabirds which had occupied them all summer. Reaching Cape Bird at the west end of Bellot Strait, McClintock climbed a 1300-ft mountain to check the sea ice conditions in Franklin Strait to the south.

At last, on her fifth attempt, the *Fox* succeeded in steaming into Bellot Strait, and preparations were made to winter in a bay they named Port Kennedy, half way to Cape Bird. The autumn was also a time for planning for the following spring's sledging expeditions. McClintock proposed to travel to the shores of King William Island and the Great Fish River, whilst Hobson explored the west coast of the Boothia Peninsula, and Allen Young travelled north up the west side of Somerset Island. They built an igloo observatory south of the ship's mooring from which regular readings of the sun's altitude were made throughout the autumn. A plate in McClintock's book *The Voyage of the Fox* shows Lieutenant Young hunched over an instrument resembling a sundial surrounded by a wall of ice-blocks and watched anxiously by a sledge dog from the doorway.

Fortunately for McClintock, Bellot Strait was an interesting area geologically, being on the junction of granite and Silurian limestone rocks. Dr Walker found nine different species of shellfish preserved as subfossils in the

4.5 McClintock reconnoitres the Bellot Strait in a rowing boat

hills between 100 and 550 feet above sea level near Port Kennedy. These, and similar finds by McClintock and earlier ones made by Parry, proved that long ago the whole Arctic archipelago had been submerged beneath the sea.

By December, animals for the larder were becoming scarce. Chilling mists from Bellot Strait and frequent gales made it too cold to leave the ship. Despite this, Christmas was attended to with great spirit. Provisions included roast venison, beer, gifts of new clay pipes for all the crew, a large cheese and even candles for the dinner table. Meanwhile 'a fierce north-westerly howled loudly through the rigging, the snow-drift rustled swiftly past, no star appeared through the oppressive gloom, and the thermometer varied between 76 and 80 degress *below the freezing point*'.

The weather had improved enough six weeks later to allow McClintock and Young to make a sledge journey to the southwest edge of the Boothia Peninsula. Each night they built a hut of snow blocks covered with their tent into which the sledge was unloaded. Once the dogs were fed, the door

> was blocked up with snow, the cooking-lamp lighted, foot-gear changed, diary written up, watches wound, sleeping bags wriggled into, pipes lighted, and the merits of the various dogs discussed until supper was ready; the supper swallowed, the upper robe or coverlet was pulled over, and then to sleep. Next morning came breakfast, a struggle to get into frozen mocassins, after which the sledges were packed, and another day's march commenced.

Half way down the peninsula, the party met a group of Eskimos, one of whom had a Royal Navy button on his clothes. He told them that the button had belonged to white men who had starved near a river where there were salmon. These, they assumed, were the survivors of Crozier's party, whose remains Dr Rae had found. McClintock offered to barter knives, files, scissors and beads (all priceless items to the Eskimos) for information. The next day the whole village of 45 people arrived at their camp. With them they brought further evidence of the Franklin expedition including six silver spoons and forks, a silver medal belonging to MacDonald who had been Franklin's assistant surgeon, a gold chain, buttons and other artefacts made of wood and iron from the wrecked ships. Petersen understood from the Eskimos that they had seen a three-masted ship crushed by the ice, and that the white men's bones were buried upon the island where they had died.

McClintock and his men returned to the *Fox* without delay to prepare teams for their sledge journeys. A gap in his diary between March and June 1859 is attributed to 'the wild sort of tent-life we lead in Arctic exploration (which) quite unfits one for such tame work as writing up a journal'. He set out on 12th April 1859 with Petersen, two sledges and 12 dogs, accompanied by Hobson's sledging team. McClintock was driving the five puppies and a small sledge which he intended to sell to the Eskimos, while Petersen took charge of the main sledge. They carried with them about 12 cwt of equipment including a tent, blanket sleeping-bags, cooking utensils, guns and ammunition, six backpacks of spare clothing, articles for barter and over 8 cwt of provisions, mainly pemmican, biscuits and tea with some boiled pork, rum and tobacco.

The teams travelled south, moving fast on smooth level ice, and reached the south-western tip of the peninsula in 16 days. Here McClintock generously sent Hobson and his team to search the west coast of King William Island where they fully expected, from the Boothia Eskimos' description, to find traces of the *Erebus* and *Terror*. McClintock continued south and across to the east side of King William Island, travelling mainly at night, when the sun was low in the sky, to avoid snow-blindness. By chance, they found an occupied Eskimo snow village where they bought six pieces of silver plate bearing the crests or initials of Franklin, Crozier and other officers in the expedition.

The Eskimos told McClintock of a wreck off the island's west coast, most of which had been removed already. There had been many books, they said, but these were destroyed by the weather when the ship finally was driven ashore. One woman told Petersen that many white men had died on their way to the Great Fish River and some were never buried. This was confirmed by what McClintock and his team found on their way south. They reached Matty Island at the mouth of the river in mid May, but found no trace of a cairn in which Franklin's men might have left a message. The island, McClintock noted, was composed mainly of grey gneiss striped with conspicuous vertical bands which lay so regularly in a north-south direction that, when crossing the island, he could navigate by them.

On 19th May they turned for home. With great relief, McClintock handed over the puppy team to Hampton, the expedition's only sickly member. In his diary he wrote:

> I shall not easily forget the trial my patience underwent during the six weeks that I drove that dog-sledge. The leader of my team, named 'Omar Pasha', was very willing, but very lame; little 'Rose' was coquettish, and fonder of being caressed than whipped; from some cause or other she had ceased growing when only a few months old; she was therefore far too small for heavy work; 'Darky' and 'Missy' were mere pups; and last of all came the two wretched starvelings, reared in the winter, 'Foxey' and 'Dolly'. Each dog had its own harnesss, formed of strips of canvas, and was attached to the sledge by a single trace of 12 feet long. None of them had ever been yoked before, and the amount of cunning and perversity they displayed to avoid both the whip and the work was quite astonishing. They bit through their traces, and hid away under the sledge, or leaped over one another's backs, so as to get into the middle of the team out of the way of my whip, until the traces became plaited up, and the dogs were almost knotted together; the consequence was that I had to halt every few minutes, pull off my mitts, and, at the risk of frozen fingers, disentangle the lines.

Despite the disadvantages of this dog team, McClintock was able to prove that sledging could be made much easier by using dogs instead of men to pull the sledges.

They stopped to search the south shores of King William Island, the actual coastline along which the survivors of Franklin's party had struggled. On a gravel ridge near the beach, they found a skeleton lying as it had fallen, on its face. The body was of a young man, apparently a mess steward. His clothes-brush and horn pocket-comb were still in the pockets of his braided blue jacket. At the western end of the strait, McClintock searched in vain for a cairn in which, he felt convinced, news of Franklin's expedition would have been left. (In these days before radio, messages left in cairns were almost the only method of communication for Arctic travellers.)

A few miles further on they discovered a cairn and a note left by Hobson's party six days before. Hobson had found nothing of the wrecks and seen no Eskimos to question but, in a cairn at Victory Point on the northwest corner of King William Island, he had found written record of the Franklin expedition. This was inscribed on one of the printed Admiralty forms usually supplied to ships to be thrown overboard in bottles in order to study current systems at sea. Upon this piece of paper, on 28th May 1846, Lieutenant Gore had recorded how *Erebus* and *Terror* had wintered at Beechey Island in 1845-6 (Gore wrote 1846-7 by mistake) after circumnavigating Cornwallis Island and returning to Barrow Strait. A party of officers and six men had visited the cairn on 24th May 1847 and recorded that they had spent the previous winter in the ice north of Victory Point. Around the edge of the page was added:

> April 25, 1848 — H.M. ships 'Terror' and 'Erebus' were deserted on the 22nd of April, 5 leagues N.N.W. of this, having been beset since 12th Sept., 1846. The officers and crews, consisting of 105 souls, under the command of Captain F.R.M. Crozier, landed here in lat. 69°37'42"N., long. 98° 41'W. Sir John Franklin died on the

11th of June, 1847; and the total loss by deaths in the expedition has been to this date 9 officers and 15 men. (Signed) F.R.M. Crozier, Captain & Senior Officer. James Fitzjames, Captain H.M.S. Erebus. and start to-morrow, 26th, for Black's Fish River.

Hobson had found quantities of clothing and other abandoned belongings around the cairn showing that, at last, the men had realised their lives were at stake on the march and superfluous baggage would endanger them further. The constant gales and fog during his journey down the west coast, Hobson added, had diminished his chance of finding the wrecks.

At the westernmost point of King William Island, McClintock and his men discovered other remains. A large ship's boat had been abandoned with more clothing. The boat had been carefully equipped for travelling up the Great Fish River, but it weighed over 8 cwt. Even mounted on its sledge it would have been an exhausting burden for men already weakened by lack of food and scurvy. The total weight of the load would have been about 14 cwt, McClintock estimated, enough for seven strong healthy men.

Lying in the boat were two corpses, a slight young man, complete with slippers, guns — one of them loaded and cocked — several books including a New Testament Bible and a copy of *The Vicar of Wakefield*, a quantity of clothes and a variety of other small possessions and tools. 'In short,' concluded McClintock, 'a quantity of weights of one description and another truly astonishing in variety, and such as, for the most part, modern sledge-travellers in these regions would consider a mere accumulation of dead weight, but slightly useful, and very likely to break down the strength of the sledge-crews.' The only remaining food was some tea and a little chocolate. The boat was about 50 miles from the position of the abandoned ships, over 60 miles from the skeletons of the steward and boy, and 150 miles from the Great Fish River. Beside another cairn at Victory Point they found more relics, including the boat's heavy cooking stoves, pickaxes, iron hoops, a copper lightning conductor and a pile of clothes four feet high.

By the time McClintock's party returned to the *Fox* on 19th June, the thawing sea ice was flooded and impassable to the sledges. The last 17 miles were travelled in an overland scramble. The dogs had become so sore-footed they had refused to leave the sledge when it was temporarily abandoned at the mouth of Bellot Strait.

McClintock and Hobson had proved, at last, the existence and position of a navigable Northwest Passage. It lay south along the east coast of Somerset Island and through Bellot Strait, then south again and round the east and south sides of King William Island to the clear water south of the huge Victoria Island. The more direct passage to the west of King William Island, McClintock concluded, nearly always would be impassable to ships because of the choking ice driven south from the channels to the northwest and northeast. These were named later McClintock Channel and Franklin Strait respectively.

Five days before McClintock's arrival, Hobson had returned to the *Fox* suffering from scurvy and unable to walk or stand without help. Another member of the ship's crew, the steward Thomas Blackwell, had died of scurvy a few days before. This was not unexpected since Blackwell had disliked preserved meat and potatoes, and had insisted on living the whole winter on salt pork.

By 25th June, McClintock was becoming worried about Young and his team who had not returned from their journey northwards, He set out with four men to look for them and to leave food. First, they returned to rescue the dog team which they found lying quietly by the abandoned sledges. 'They had attacked the pemmican, and devoured a small quantity which was not secured in tin, also some blubber, some leather straps, and a gull that I had shot for a specimen; but they had not apparently relished the biscuit', McClintock noted.

Climbing up to look out from the cliff tops

when they reached Cape Bird, McClintock was amazed and delighted to find some brent geese nesting on an accessible ledge. There was no sign of Young's party and McClintock sent the men back to the ship while he and Thompson continued up the coast with the dog team. The same day, to the joy and relief of all concerned, they met Young and his men heading south towards the *Fox*.

Life gradually became easier as the summer progressed. Anderson and the Eskimos shot seabirds, but were frustrated in their attempts to find or reach the nests of the crafty eider ducks, geese and gulls. One of the men appeared with a trout weighing 2lb which had been brought ashore by a glaucous gull. The bird was immediately challenged by an Arctic fox, and the man had scared the contestants off and grabbed the prize. As well as all the fresh meat they could eat, they varied their diet with mustard and cress, the only crop to grow on board the *Fox*, sorrel leaves and the roots of a plant with flowers which, McClintock said, resembled snapdragons. This species has the unappetising name of hairy lousewort but a pretty pink flower, and grows commonly in damp places on the Arctic tundra. McClintock added to his plankton collection which had been started from the stomach contents of a seal shot the previous summer. When he returned home, the collection was examined by the Rev. Eugene O'Meara of the Royal Dublin Society and found to contain 85 species of animals.

The *Fox* was cleaned and repainted in preparation for the homeward voyage. The ship's engines had been dismantled during the winter, but the engineer, George Brands, had died from a stroke on 6th November. No one knew what should be done with the engines so McClintock, with his usual flair for technical matters, settled down to rebuild them. On 1st August, the pack-ice began to move out of Port Kennedy bay taking the *Fox* with it. The engines were started up and run at full steam until McClintock was satisfied they were serviceable. Ten days later they were under way, and, after minor engine trouble, they moved back up the coast of Somerset Island.

McClintock drew up a list of the game they had shot during their two years in the pack-ice. The first winter, 1857–8, they had killed 2 polar bears, 73 seals, 38 dovekies and an Arctic fox. During the 11 months at Port Kennedy, they had shot and eaten 2 more bears, 8 reindeer, 9 Arctic hares, 19 Arctic foxes, 82 ptarmigan, 98 wildfowl (ducks, geese and gulls) and 18 seals. Between them, the three sledging expeditions had added 800 miles of new coastline to the map of the Arctic. The expedition, everyone agreed, had been a great success.

By the end of August they were steaming through 'calm, warm, lovely weather' to moor once again in Godhavn (Godthaab) harbour on the Greenland coast. Ten days later they rounded Cape Farewell at the southern tip of Greenland, and a week of favourable gales sped the *Fox* on her way towards Land's End. They arrived at Portsmouth on 20th September 1859 after over two years at sea.

The Admiralty was delighted with McClintock's discoveries. Hobson was promoted to commander, and McClintock received a knighthood and £1500. The whole of Britain welcomed the *Fox* and her crew, and her homecoming took on the air of a happy ending to a story. In 1869, Lady Franklin was presented with the Royal Geographic Society's Gold Medal in commemoration of her husband's discoveries. A statue of John Franklin was erected in Waterloo Place, London, and a bust of him placed in Westminster Abbey. Its inscription bore the now famous lines by Tennyson, who was married to Franklin's niece:

> Not here! the white North hath thy bones, and thou,
> Heroic sailor soul,
> Art passing on thy happier voyage now
> Towards no earthly pole.

What touched McClintock most about his return was, as he put it,

the purchase of a gold chronometer, for

presentation to me, [which] was the first use the men made of their earnings; as long as I live it will remind me of that perfect harmony, that mutual esteem and goodwill, which made our ship's company a happy little community, and contributed materially to the success of the expedition.

The expedition had succeeded mainly because, as his biographer Markham says, McClintock had all the elements of a good explorer: the naval training, the personal qualities and the ambition. To this McClintock's friend, the Rev. William Alexander, Archbishop of Armagh, added:

Usually the Admiral was reserved and somewhat indisposed to talk, but the approach of an emergency, possibly not without indications of danger, seemed to inspire him with the lofty touch of exhilaration (sic) which is the peculiar gift of the bravest alone. His face lit up with animation, his words came with more than usual readiness and cheerfulness of tone.

McClintock prepared his diaries for publication and in 1860 *The Voyage of the 'Fox' in the Arctic Seas. A Narrative of the Discovery of the Fate of Sir John Franklin* appeared. The book was an immediate success. Ten thousand copies were printed, and 7,000 of these reportedly sold before the book was published. Eventually it ran to seven editions. Samuel Haughton, who examined McClintock's fossil collection, added an appendix to the book. The specimens were deposited in the museum of the Royal Dublin Society and are now housed in the National Museum in Dublin. They were, Haughton concluded, 'a more extensive and better collection of Arctic rocks and fossils than is to be found in any other museum in Europe.'

McClintock's Arctic expeditions between 1848 and 1859 had been successful in a number of different ways. In addition to exploring enormous areas of previously unknown Arctic terrain, discovering what had happened to Franklin's expedition, and proving new techniques of long-distance sledging, McClintock made considerable contributions to the knowledge of zoology, botany, geology, hydrology and meteorology. He was an intelligent and accurate observer and a meticulous record keeper. He had made a series of sun sightings wherever he went, calculating his latitude by its meridian altitude and his longitude by chronometer readings.

McClintock also kept many other records which are rarely referred to in his diary, but described in a letter to the Royal Society which was published in its *Proceedings* for 1859–60. These included the observations made in the magnetic observatory built of ice blocks near the *Fox*'s winter moorings, and meteorological records including measurement of the amount of ozone in the atmosphere at Port Kennedy. Atmospheric electricity was measured with an 'electroscope' and the effects of the aurora borealis upon this were noted. He conducted experiments on the differences in specific gravity and temperature of the sea water from various depths, and kept a series of tidal records.

McClintock's collection of botanical and zoological specimens included the first ivory gulls' eggs ever to reach Europe (found on Prince Patrick Island). His plants and seaweeds, which had to be abandoned on board the *Investigator* in 1853, eventually were returned to him by Dr Krabbe, second master of the *Assistance*, who visited the ship the following year. McClintock's biographer lists 60 species in his collection of Arctic plants, and 31 species of zoological fossils including the new ammonite and trilobite already mentioned and a number of apparently new molluscs. McClintock was elected a Fellow of the prestigious Royal Society in 1865.

Within eight months of his return from the Arctic, McClintock was at sea once more, and his commands in the following years took him to Newfoundland, the Mediterranean including Cairo and the pyramids at Gizeh, the North Sea and Barbados. In October 1865, Sir Leopold McClintock was appointed Commodore and given an official residence at Port Royal on Jamaica. Here, despite recurring bouts of yellow fever, he found time to complete writing up his account of Arctic sledging techniques and equipment.

He also continued to collect specimens while he was in the West Indies. Fish and marine invertebrates from Barbados and Jamaica were despatched to the Royal Dublin Society, and with them 'two skins of a rare petrel known as a "Blue Mountain Duck".' This is actually a seabird, called a black-capped or Jamaica petrel which is now thought to be extinct. The Jamaica petrel was found only on Jamaica where it nested in burrows and, like many of the other gadfly petrels to which it is related, it owed its demise to animals like rats and mongooses which were introduced onto the island by man. These efficient predators caused the rapid depletion of petrel numbers during the eighteenth and nineteenth centuries, until only a very few birds remained living high on inaccessible mountainsides in the middle of Jamaica. The last know specimens of the Jamaica petrel were collected only a few years after McClintock left Jamaica in the summer of 1868.

On his return from the West Indies, McClintock went to Ireland where he stood as the British Conservative party's candidate in a local election at Drogheda, then a part of Britain. For the first time in his 49 years, he was able to develop his private life. During the electioneering he met and fell in love with Miss Annette Dunlop of Monasterboice, County Louth. In the spring of 1869, having failed to obtain a seat in Parliament, McClintock returned to London to serve, among other things, on the Council of the Royal Geographic Society to which he had recently been elected. On 12th October 1870, McClintock was married to Annette Dunlop in Ireland, and afterwards the couple moved to Eaton Terrace in London. Two years later McClintock was appointed Admiral Superintendent of Portsmouth Dockyard.

During 1874 and 1875, McClintock's job involved him in yet another Arctic expedition. He was put in charge of equipping and advising the British Arctic expedition lead by Captain Nares. Both the expedition's ships were strengthened and adapted to McClintock's instructions, and fitted with boats that could be used for whaling or as sledges. When the expedition left Portsmouth in May 1875, the leading ship flew a silk flag, made by Lady McClintock, bearing a Union Jack on one side, with the McClintock crest and a rose, thistle and shamrock on the other side.

4.6 Ivory gulls nest only in the high Arctic, and the first clutch of their eggs ever brought to Europe was collected by Leopold McClintock (Courtesy of the Alexander Library, EGI, Oxford)

4.7 Admiralty House on Clarence Hill in Bermuda, one of the McClintock family homes

The term of office in Portsmouth came to an end in 1879, and the McClintocks moved back to London. By this time McClintock had reached the rank of Vice Admiral and the following November he was appointed Commander-in-Charge of the Navy's North American and West Indian station. A month later he and his wife, his three sons and a daughter sailed for Bermuda. It provided him with a welcome opportunity to renew his interest in natural history. After three happy years, the family returned, with the addition of a second daughter, to England where McClintock faced his approaching retirement. 'A timely intervention' induced a senior Admiral in the Navy to retire voluntarily just in time to enable McClintock to be promoted to full Admiral before his own retirement.

In the years that followed, McClintock continued to contribute his vast knowledge and experience to scientific circles. He served on various councils and, in 1884, he was elected an Elder Brother of Trinity House, the British lighthouse service. Unlike many retired specialists, he retained a generous open-mindedness to new ideas, as this account shows:

In 1892, when Nansen presented to the Royal Geographic Society his plan to drift the *Fram* in the Arctic ocean pack[-ice], he received a discourageing reception generally. McClintock, however, considered it 'the most adventurous programme ever brought under the notice of the Royal Geographic Society'. It bespeaks the typical generosity of one of the true gentlemen of Arctic exploration.

McClintock kept up his work into his eighties, despite bad health and increasingly poor eye sight. On the fiftieth anniversary of the sailing of the *Fox*, he attended a luncheon with his family and surviving 'messmates' from his Arctic days. These included Sir Allen Young from the *Fox's* crew and his biographer, Sir Clements Markham.

A few months later, on 17th November 1907 and at the age of 88, Sir Leopold McClintock died. The mourners at his funeral included representatives of the Royal Family and many other dignitaries. Among the wreaths was one from the Royal Greenland Company, current owners of the *Fox*, herself 52 years old. Fittingly, an alabaster slab commemorating Sir Leopold McClintock was placed under the bust of Franklin in Westminster Abbey. In addition to all his other talents, it states, McClintock had been 'a great seaman, a great explorer and, in the highest sense, a great man.'

Henry Bates's journeys in Brazil, 1848–59

CHAPTER FIVE
Eleven Years on the Amazon

5.1 Scarlet-faced monkeys or red-faced ukaris, the only group of South American monkeys without long prehensile tails, and the parauacu or monk saki

On 26th May 1848 a small trading vessel named the *Mischief* arrived at the mouth of Brazil's greatest river, the Amazon. She anchored in deep water, six miles off Salinas, or Salinopolis as it is known today, and raised a pilot flag. The village, separated from her by sandy shallows, had been founded as a Jesuit missionary settlement, and now provided pilots for the River Pará, the southernmost of the two huge outlets of the Amazon waters into the Atlantic. On board the *Mischief*, two passengers scanned the distant coastline eagerly through the captain's telescope. Bare sand dunes lay to the east. Westwards, a long line of forest reaching down to the water's edge stretched away through the heat to the river's estuary. This, at last, was the 'great primaeval forest' which covered the country for two thousand miles west up to the foot of the Andes.

The two men had embarked at Liverpool exactly one month earlier, and had had a swift and uneventful passage from England, once the storms and sickness of the Bay of Biscay had been endured. Although they shared a common, almost fanatical, interest in natural history, especially in insects, they were travelling companions rather than close friends.

The younger of the two, Henry Walter

Bates, had been born in Leicester 23 years earlier, the son of a hosier in the small midland town. His mother, described as 'of sweet, loving and unselfish nature but of feeble constitution', gave birth to three more sons after Henry. The oldest child must have delighted his parents by studying fervently in school, and winning prizes for Greek and Latin. Soon he became just as keenly interested in insects, and at the age of only 18 his first paper, on beetles, entitled 'Notes on Coleopterous insects frequenting damp places', was published in the new monthly journal *The Zoologist. A Popular Miscellany of Natural History*. On finishing his formal education in 1838, Bates worked first as an apprentice in his father's business, and then at Allsopp's brewery in nearby Burton-on-Trent.

Two events in 1847 brought him closer to fulfilling his ambition to travel. For three years Bates had been corresponding with a fellow-entomologist called Alfred Wallace, and in 1847 he paid a visit to Wallace's home in Neath, South Wales. In the same year, an American named W. H. Edwards published a brief account of his journeys in South America called *A voyage up the Amazon* which both men read. Before long they had agreed to travel the Amazon themselves. They were to collect specimens of the wildlife they found there, and as Wallace, who was to be with Charles Darwin the formulator of the theory of evolution by natural selection, told Bates in a letter, they would study 'towards solving the problem of the origin of species'.

Alfred Russell Wallace, Bates's companion and his senior by two years, was a tall, athletically built man with dark wavy hair, side whiskers and a neat beard, and deep-set blue eyes almost hidden behind his spectacles. He had been born in Monmouthshire in South Wales, and lived his early life in a cottage beside the River Usk where he soon learnt to catch lampreys. His only education was a brief spell at Hereford grammar school, and in 1836 family circumstances forced him to go and live with his elder brother, William, who was a land surveyor. Together they travelled the country from the windy fens of Bedfordshire to the mountains of South Wales surveying routes for the new railways.

By 1840, William had died after catching a chill whilst travelling, ironically enough, by train; and Alfred had become deeply interested in natural history. He was a dedicated collector of local plants and insects when, four years later, while teaching English in the Collegiate School in Leicester, he happened to meet Henry Bates in the local library. The two men immediately discovered their common interest. Wallace was very impressed by Bates's collection of butterflies and beetles, and even more astonished to hear that, in addition to these species collected around the town, thousands more existed and many others were still to be discovered.

The *Mischief* sailed into the Pará river on 27th May 1848, although the estuary was so broad that only the calmness and discoloration of the water showed them they were no longer at sea. On the eastern shore, ten miles away, a continuous line of forest was broken only by the occasional fishing villages. At sunrise the following day they arrived at the city of Pará (now called Belém). The river was still 20 miles across although they were more than 60 miles from the sea. The air felt oppressively humid, sheet lightning flickered about the horizon.

Pará was a city of about 15,000 people built on a low river bank and hemmed in by the constantly encroaching forest. Its buildings were mostly whitewashed, and had red tiled roofs topped here and there by towers. Pará had been founded in 1615, and two centuries later it had grown into a thriving port with a community of merchants and tradesmen, a population of 24,500. In 1819, a smallpox epidemic swept through the city killing many of the Indians and Europeans, and it had never recovered its former size. The day Bates and Wallace arrived in Pará a religious festival was being celebrated in the city. While they were absorbed in the sight of

> the broad glassy river, studded with richly wooded islands, resting like floating gardens on the water, the crowd of native canoes with the motley clusters of half-

naked people of all shades in colour of skin ... the bells of the numerous churches began a lively peal, trumpets sounded, sky-rockets whizzed in the air, and guns boomed from the forts and the shipping. It was a pleasant introduction to our new home.

The two men moved into a house belonging to Miller, the owner of the *Mischief*, which was situated right at the edge of the city and close to the forest. The first evening they walked through the suffocatingly hot, moist and mouldy air. Bates, like so many new arrivals in the tropics before and since, was never to forget those searing first impressions. He was overwhelmed by the busy streets, the people, the fragrant beauty of the fruit trees and climbing plants in the midst of the slums, and above all the unbelievable noise of cicadas, frogs and toads in the twilight. 'This uproar of life', he wrote, 'I afterwards found never wholly ceased, day or night.' When at last he returned to England, the 'death-like stillness of summer days' seemed as strange to him as the cacophony which accompanied his first walk through Pará.

Miller's suburban house was an excellent starting point for the collectors. Close by was a swamp drained by the tides and, in the forest, trees towered to enormous heights, festooned with climbing parasitic plants with shining leaves. The canopy echoed with the calls of birds which were tantalisingly difficult to see, raucous and brightly coloured parakeets, darting flycatchers, and velvety black and yellow orioles called troupials whose purse-like woven nests hung down three feet from the tips of branches. Around the house were wrens which, Bates decided, sang like English robins and with them, living just like sparrows, were deep crimson silver-beaked tanagers. Lizards of many kinds were everywhere and gekos, 'very repulsive in appearance' according to Bates, lived inside the house, as did bats which flitted about the rafters of the ceilingless rooms. The most conspicuous insects were the butterflies. After the fewer than 70 common British species, their diversity was amazing. Among them were swallowtails, brimstones, vanessids and whites, and by the end of the first three weeks they had collected 150 different species each.

The insects which most interested Bates were the tropical ants. They marched about the suburbs in broad columns, with some individuals an inch long. Mounds of freshly excavated earth up to six feet across, had puzzled Bates and Wallace during their early walks through the forest. Now they discovered these were made by sauba ants whose habit of clipping and carrying off huge quantities of leaves and other greenstuffs with which to thatch their nests extensively damaged many of the trees.

Bates refused to believe the local stories of ants taking food from the houses. Some years later he was staying in an Indian village on the River Tapajos where food was very scarce, when the servant awoke him in the middle of the night claiming that rats were stealing the precious farinha, a granular mixture of tapioca starch and woody fibre from the tapioca plant. Instead, Bates found that sauba ants were systematically raiding the store baskets, passing grains between them. When less drastic preventative measures had failed, Bates 'was then obliged to lay trails of gunpowder along their lines, and blow them up'!

After a couple of weeks, Bates and Wallace moved to a house in the village of Nazareth, about two miles from the city but almost completely surrounded by virgin forest. Here they settled happily into a routine of collecting specimens. They rose in the cool clear-skied dawn and watched or caught birds and other animals until breakfast at 10 a.m. Then, in the increasing heat, they collected insects until the sea breeze which always arose in the mornings precipitated its daily torrential shower. The day's work kept their evenings busy examining, discussing, identifying and recording the specimens; then preparing and preserving them; and finally storing them carefully away in jars and boxes for dispatch to England.

The two men started out on foot on 23rd June to visit a saw mill and rice-cleaning mills at Magoary, 12 miles away. The manager of the mills was a Canadian called Leavens who acted as their guide during the visit. It proved such a success that they returned in July equipped 'with hammocks, nets and boxes', and stayed ten days.

They had taken a while to become accustomed to working in virgin forest which was very different from the secondary growth in the forests around Pará. At first, it felt to Bates like an 'inhospitable wilderness', and the birds and mammals seemed disappointingly scarce. Soon he realised that was only because they were shy of man, and before long he was able to spot even the elusive monkeys swinging in the high canopy by their tails. There seemed to be no limit to his collection of birds which now included toucans, parrots, hummingbirds, ant-thrushes, manakins and a whole range of

5.2 Bates hunting toucans in the Amazonian rain forest

waterbirds from the lily-covered lakes near the mills. They returned to Pará on foot, leaving their luggage and collections to travel by canoe. Two months' work had secured over 400 kinds of butterflies, 450 species of beetles and about the same number of other insects.

Leavens offered to take the naturalists with him on a trip up the River Tocantins, the largest tributary of the River Pará, in search of cedar wood suitable for his mill. The three men started out on 26th August in a 25 ft canoe with two palm-thatched shelters and two masts. It was loaded down with a crew of four, guns and ammunition, specimen boxes, three months' supply of farinha and other dried foods, and several bushels of copper coins which were the only recognised currency on the Tocantins.

Sometimes they slept cramped in the canoe, waking to the calls of waterbirds along the river. More often they stayed with local traders or plantation owners along their route with whom Leavens was acquainted, or to whom they had letters of introduction. Most of the villages they passed through were prospering on a trade of Brazil nuts, cocoa, rubber and cotton. At one point they were delayed by the desertion of most of their crew and the difficulty of replacing them, but this hardly bothered Bates and Wallace who were busy collecting. Hoatzins were plentiful in flocks along the river but they rarely came to the ground and were difficult to shoot for specimens. These bronze-plumaged, pheasant-like birds with their bare blue faces feed entirely on leaves but are also excellent swimmers. Young hoatzins are unique among birds in possessing claws on the joints of their wings which they use to help them climb back into the nest or from the water onto overhanging branches.

At the home of one of the Brazilian planters above the town of Cametá, Bates saw his first cotinga, a 'sky-blue chatterer ... on the topmost bough of a very lofty tree and completely out of range of an ordinary fowling piece'. No wonder he wanted a specimen so badly for cotingas are among the gaudiest and most beautiful of all South American birds.

The travellers reached Baiao, about 100 miles from the river's mouth, at the beginning of September. The village was built on so steep a bank that they had to climb up to it from the canoe by a ladder. By now they were out of the low-lying Amazonian floodplain, and to their relief the air was clearer and drier. As usual, they skinned and prepared specimens in the open window or on the balcony of their house, and a crowd of incredulous natives gathered. 'Oh, the patience of the whites!,' they exclaimed as they watched the delicate work, 'Does he take all the meat out?', 'Well, I never' and 'Look, he makes the eyes of cotton!'.

A week later they arrived at the village of Patos where they made a detour from the main river with Leavens to search for cedar trees. In the deep forest they saw but failed to collect a magnificent hyacinthine macaw. This rare parrot is over three feet in length with splendid purplish blue feathers. Bates watched it feeding on nuts which he himself found almost too hard to break using all his strength and a hammer, but which the parrot effortlessly crushed to a pulp in its beak.

By 16th September, the party had reached Arroyos, over 120 miles upriver and the end of their outward journey. The bed of the Tocantins was a mile and a quarter wide at this point, and blocked by a roaring cataract. Their passage downstream was accomplished in only ten days. In the sea-like expanse of the river mouth, Bates saw freshwater dolphins. Dolphins are more familiar as marine mammals although several species are found in the larger rivers of South America, and another species lives in the Ganges River in India. Bates described two species of dolphin from the Tocantins and knew of a third species on the upper Amazon.

A couple of months after his return to Pará, Bates decided to make a trip alone to the coast of Carnapijo, a peninsula extending towards the huge island of Marajó which separates the estuary of the River Pará from that of the main Amazon. He travelled on a small trading vessel, leaving Pará on 7th December. 'We were 13 persons aboard,' he wrote, 'the cabo [captain], his pretty mulatto mistress, the pilot and 5 Indian canoemen, 3 young mamelucos [tailor apprentices who were making a holiday trip to Cameta], a runaway slave heavily chained, and myself.'

He was to stay with a Scottish planter at Caripi, an area which already had, as Bates puts it, 'quite a reputation for the number and beauty of the birds and insects found there.' He soon resumed his collecting routine, and may have been glad to be relieved of competition for specimens with Wallace.

Caripi did not disappoint him. Even in the garden, numerous hummingbirds fed at the orange blossom, and several times he shot a hummingbird hawk moth mistaking its size and behaviour for that of a bird. He met a German settler, living nearby with his family, who was an entomologist and a reckless sportsman. He and Bates made a hunting trip to the islands at the tip of Carnapijo. By the time he returned to Pará on 12th February 1849, Bates had succeeded in amassing what he proudly describes as 'a very large collection of beautiful and curious insects, amounting altogether to about twelve hundred species'. He concluded that the abundant rotting wood in the new forest clearings made by the Caripi natives must have been responsible for attracting an exceptional variety and great numbers of beetles and other insects. His spirits and confidence were boosted even further by a letter from his agent in London reporting that his specimens were selling well.

Bates next revisited Cameta on the lower reaches of the Tocantins River. He set out on 8th June, and at about the same time Wallace left to explore the Guama and Caprim rivers which entered the Pará. Bates travelled on a trading vessel of 30 tons which was so full of cargo that its passengers had to sleep on deck. Here, as no one could sleep anyway, they were serenaded by John Mendez, the musical pilot. 'He had on board a wire guitar or viola, as it is here called,' explains Bates, 'and in the bright moonlit nights, as we lay at anchor hour after hour waiting for the tide, he enlivened us all with songs and music.'

He was to find that the effect of the Atlantic tides was felt as far up the Amazon as Barra, at its junction with the Rio Negro, and also on most of its tributaries.

The plantations around Cameta were rich collecting grounds for Bates. In a letter to the *Zoologist* he describes how he set out at 9 a.m. each morning:

> over my left shoulder slings my double-barrelled gun... In my right hand I take my net; on my left side is suspended a leathern bag with two pockets, one for my insect box, the other for powder and two sorts of shot; on my right side hangs my 'game bag,' an ornamental affair, with red leather trappings and thongs to hang lizards, snakes, frogs, or large birds; one small pocket in this bag contains my caps; another, papers for wrapping up the delicate birds; others for wads, cotton, box of powdered plaister (*sic*), and a box with damped cork for microlepidoptera; to my shirt is pinned my pincushion, with six sizes of pins.

Bates was particularly intrigued by the giant hairy spiders of the genus *Mygale*. He found one web in a tree crevice in which the spider had caught two small finches. In another place he saw some Indian children leading a spider 'about the house as they would a dog' by a light cord attached around its 'waist'.

The return journey to Pará was full of adventures and typical of Amazon travel. The *Santa Rosa*, a two-masted covered canoe of the type known as a cuberta, in which all Bates's luggage including his specimens had been stowed, departed unexpectedly leaving him behind. He anxiously sought a lift with an Englishman who was to return to Pará in a small boat, leaving about midnight. But, as Bates recounts, 'about seven, the night became intensely dark, and a terrific squall of wind burst forth, which made the loose tiles fly over the housetops; to this suceeded lightning and stupendous claps of thunder, both nearly simultaneous.' He began to doubt whether he would ever see his possessions again.

'At midnight, when we embarked, all was as calm as though a ruffle had never disturbed

5.3 This species of bird-eating spider, found by Bates near Cametá, can reach two inches across the body with legs up to five inches long

the air, forest or river.' The small canoe sped along through the night 'to the rhythmic paddling of four stout youths' and, by dawn, it had caught up with the *Santa Rosa*, now becalmed and at anchor off a midstream island. Preferring to complete the three days' journey with his precious specimens, Bates thanked the Englishman and joined the *Santa Rosa*. He was slightly alarmed to find that the cuberta was heavily laden and leaking copiously. The crew were all in the water, diving down to feel for holes in the hull which they were blocking ineffectively with rags and clay. Meanwhile an old negro was patiently baling the water out of the hold.

The journey proceeded surprisingly smoothly until they were within sight of Pará when the breeze suddenly stiffened and caused the leaks in the hull to break out again. The captain, Senhor Machado, with whom Bates had become rather friendly, ordered another sail to be raised in the hope of making the shore two miles away before the cuberta sank. In the next gust of wind, the rigging gave way and the boom and sails collapsed onto the deck.

> We were obliged to have recourse to oars, and as soon as we were near land, fearing that the crazy vessel would sink before reaching port, I begged Senhor Machado ... to send me ashore in the boat, with the more precious portion of my collection.

Undaunted by these experiences, Bates departed for the upper Amazon a couple of months later. Although this was to be one of his longest journeys, he had no choice but to travel by sailing boat as there was no steamer service on the Amazon until 1853, still four years in the future. On this occasion, the boat was a 40 ton schooner, with a crew of 12, which belonged to a local trader. Sailing close to the forested river banks broke the monotony of the slow journey upstream. The wall of trees seemed alive with birds, especially with gaudy parrots. 'We saw a flock of scarlet and blue macaws feeding on the fruits of a bacaba palm, and looking like a cluster of flaunting banners beneath its dark-green crown', wrote Bates in his diary.

That night there was roasted alligator for

5.4 Anteaters were common in the South American forests, and both Bates and Charles Waterton reported seeing them

supper, and the next day they entered the main stream of the Amazon River. Its 20 mile width was divided up by a large island and sand banks on which Bates saw noisy flocks of gulls and terns nesting. Although he does not identify them, these were probably grey-hooded gulls and large-billed terns which are the commonest river breeding species in South America. Both form colonies wherever low water levels expose banks and islands, and provide them with an opportunity to nest.

When they reached Obydos (or Obidos), Bates decided to stay behind, and he remained there for a month hunting for specimens. The forest was full of monkeys including spider monkeys of which the local Indians were very fond. They would shoot a monkey from its tree using a blowpipe and poisoned dart; then resuscitate it and keep it as a pet. The Indian women would even suckle the young monkeys, and they made good pets, following their owners everywhere like a dog. There were also enormous numbers of butterflies in the area including those of the genus *Morpho* which were 6–8 in. in wingspan.

Bates set out upstream again on an overloaded cuberta for Barra de Rio Negro, now called Manaos. The journey was made particularly memorable for him by the forest and waterbirds along the route, and by a violent storm. They reached Barra in late January 1850, after two months of travel. There Bates was reunited with Wallace, and met Wallace's brother who had recently arrived from England. They settled down to pass the worst part of the wet season in the small community of foreigners in Barra. At this time of the year, the land between the Amazon and the Rio Negro, like much of that on the lower Amazon above Santarem, flooded into a huge swamp which filled with birds and other animals, insects, and wonderful plants and flowers. The Barra community included an Irish trader, three Germans and a particularly humorous American who in spite of being a deaf mute had travelled extensively in Peru, Chile and Brazil. He 'was a fund of constant amusement both for the Brazilians and ourselves', wrote Bates.

In the spring of 1850, Bates and Wallace separated again. Wallace explored the Rio Negro and the Uapes River, one of its tributaries, and moved on to the upper reaches of the Orinoco before returning to England in 1852. Unfortunately, the *Helen*, on which he started his homeward voyage, caught fire in mid-Atlantic, and most of his live animals and specimens were destroyed. Wallace and the crew spent ten days in the boats before being rescued. This disaster alone would have spoilt the South American trip for Wallace but, in addition, his brother had died from yellow fever on his way back to England in 1850. In the autumn of 1853 Wallace published his account of the expedition, *A narrative of travels on the Amazon and the Rio Negro*. Soon afterwards he departed for Malysia and Indonesia where he was later to formulate the most controversial and exciting ideas of his life's work. These led him to a theory of evolution by means of the process of natural selection at the same time that Charles Darwin in England was coming to identical conclusions.

Bates, meanwhile, moved on, to the upper Amazon or Solimoens. He made the small town of Ega, now known as Tefe, his base for the next seven years. His ascent to Ega was made in a cuberta which had brought a cargo of turtle oil downstream to Barra. The trade winds reached inland as far as Barra, almost 1000 miles from the Atlantic, but without them the upper Amazon felt stagnant and sultry to Bates. Each day of the journey he added to his collection of birds, reptiles, insects and shells, but the low wooded banks and wide plains through which they travelled seemed interminable. The fishermen brought him pieces of pumice that had floated downstream from the great volcanoes of the Andes, still almost 1000 miles to the west.

He settled in a white-washed house in Ega and spent month after month collecting, drying and preserving specimens with his small reference library carefully arranged on wooden boxes, and drying cages slung from the ceiling on ropes smeared with bitter oil to keep out ants and rodents.

5.5 Turtle meat, eggs and oil provided a means of subsistence for many Amazon communities in the nineteenth century

Bates soon discovered that Ega subsisted on the river turtles. These grew to an unusually large size, sometimes three feet long and only slightly less across. They formed the basis of the region's main export and the principal local food. Turtles were stored alive during the dry season in little ponds behind most of the houses in the town. Bates often accompanied the local people to the river sandbanks where the turtles bred, and helped in the harvest of their eggs from which the valuable oil was extracted. On one occasion, whilst camping on an island bank, Bates's small dog gave the alarm when the camp was attacked by alligators in the middle of the night. He also visited the traditional pools, deep in the forest, where the young turtles lived. While he was there he found and collected a rare umbrella bird with its drooping crest of silky, hair-like feathers. Its huge wattle at the base of the neck was connected, so Bates discovered on dissection, to its windpipe. 'To which the bird doubtless owes its singularly deep, loud and long-sustained fluty note', he concluded.

Much as he enjoyed the collecting around Ega, Bates suffered from loneliness and a great sense of isolation. The adverse trade winds lower on the river prevented the arrival of boats bringing him letters from home and much needed funds. By the middle of 1851, he described himself as being 'shoeless, and, through robbery by a servant, penniless'. He decided that he must return to England, and he set off downstream to Para.

He made the 1400 mile journey riding the currents of the rainy season aboard a schooner heavily loaded down with turtle oil. He found the city of Pará had been devastated the previous year by two epidemics: yellow fever affecting mainly the whites, followed by smallpox which attacked the Indians, negroes and people of mixed race. No sooner had he arrived in the unfortunate city than Bates caught yellow fever, incredibly the first illness he had had since coming to Brazil. Although he recovered easily from the attack, his health was never so strong again.

Good news awaited him at Pará. The collections made at Barra and Ega which he had already shipped to England had earned him nearly twice as much as he had expected and, better still, one of the new species of insects he had found at Ega had been named *Callithea batesii* in his honour. Refreshed by the interest at home in his work, and by the welcome he received from friends in Pará,

Bates decided to stay on in Brazil. He arranged shipping for his latest consignment of specimens, and restocked for his second and longest journey into the interior.

By the end of that year, he was back at Santarem, and the following June (1852), he made an expedition up the River Tapajos. It was the hot season. The fire ants, which had plagued the party near the river's mouth, were replaced in its upper reaches by mosquitoes and other biting insects. Despite the collecting, Bates suffered from the intense heat. Although there were few birds along the river, he found some new species of fish and reptiles. The discomfort and frustrations of the journey were compensated for by a visit to a Tushaua Indian settlement near the rapids, 70 miles from the river mouth, which marked the limit of their ascent. Bates was completely charmed and intrigued by their peaceful existence and close-knit society. The descent of the Tapajos in September, the height of the dry season, was made especially hazardous by the lack of dependable currents and the shallow water. The strong daytime winds forced them to travel mainly by night and it was with considerable relief that they arrived safely at Santarem on 7th October.

Bates spent most of the following year collecting in and around the small town of Villa Nova above Santarem. Then he moved westwards again to Ega where he remained for the next five years. In 1857 his excursions took him as far upstream on the Amazon as São Paulo de Olivença, almost on the modern Colombian border and 1800 miles from Pará. After a stay of four months in São Paulo, he had a serious attack of fever which left him 'with shattered health and dampened enthusiasm'. He had been in South America for nearly ten years, mostly alone. It was time to go home.

When the steamer arrived at São Paulo in January 1858, his friends on board were shocked at Bates's appearance and low spirits, and they urged him to return to Ega with them. The steamer called in again on its downward journey the following month, and Bates took the opportunity to leave. Exactly a year later (for even going home could not be hurried!), he set out from Ega for the last time and travelled down to Pará en route for England.

Bates found the city had been much improved in the previous seven and a half years. The adjacent swamps had been drained, buildings and roads repaired, and the population had risen to over 20,000 with the arrival of many Portuguese, Madeiran and German immigrants. There were also plenty of changes he did not approve of, as he points out.

> The mantle of shrubs, bushes and creeping plants which formerly when the suburbs were undisturbed by axe or spade, had been left free to arrange itself in rich, full and smooth sheets and masses over the forest borders, had been nearly all cut away, and troops of labourers were still employed cutting ugly muddy roads for carts and cattle, through the once clean and lonely woods.

What would Bates have said if he could have known that these changes were to continue on an enormous scale throughout the rain forests of Brazil for at least the next century and a quarter?

He did not look forward to his return to Europe, as he put it:

> to live again amidst these dull scenes I was quitting a country of perpetual summer, where my life had been spent like that of three quarters of the people gipsy fashion, on the endless streams or in boundless forests.

In June 1859, Bates embarked for New York on board the *Frederick Demming*. Learning from Wallace's experience, he had been careful to divide his collection into three parts which were sent home on separate ships. This was his saddest hour. 'On the evening of the third of June, I took a last view of the glorious forest for which I had so much love, and to the exploration of which I had devoted so many years', he wrote.

Three days later, now 400 miles out to sea, among the flotsam around the ship he saw fruits from the ubussu palms found all along the banks of the great River Amazon.

Bates must have returned to English life more easily than he had anticipated for, in January 1861, still at the age of only 35, he married Sarah Ann Mason of Leicester whom he had known since his childhood. They had three sons and two daughters.

Bates's reputation as an entomologist had preceded him to England with his shipments of specimens. During his 11 years in Brazil, he had gathered hundreds of thousands of specimens representing an incredible 14,000 different insect species and over 700 species of other animals, including many previously unknown to science. Before his visit to South America, little was known about the fauna of the upper Amazon, apart from a small collection made by two Germans, named Spix and Martius, who had visited Ega. Bates's specimens from the Pará area showed its fauna to be more like that of the coasts to the north than the rest of the lower Amazon.

Through his meticulous observations of butterflies, Bates had been able to propose a theory of mimicry as a means of avoiding predation. Henry Bates is best remembered by modern biologists for his discovery of this phenomenon which is now known as Batesian Mimicry. He noticed, for example, that among the Brazilian butterflies in the family Ithomidae, many of those in the genus *Leptalis* closely resembled species in the genus *Ithomia*, both in appearance and behaviour. He concluded that varieties of *Leptalis* which were least like *Ithomia* were quickly destroyed by predators. However, he was unable to explain exactly how mimicry could confer such an advantage to *Leptalis*.

Soon after his return to England, Bates presented a paper on mimicry in insects to the Linnean Society in London. He was delighted to receive an enthusiastic letter from Charles Darwin in November 1862 which fully supported his theory. Mimicry has been demonstrated in many different insects throughout the world, but it was Charles Darwin who first explained how it works. Among the Amazonian butterflies, for example, many if not all the species of *Ithomia* are distasteful to predators such as birds or lizards which catch them. The butterflies are so disgusting in fact that the predator, having tried to eat one once in its life, immediately learns never to take a butterfly like that again. Butterflies in the genus *Leptalis*, although actually palatable to predators, gain protection by mimicking the distasteful *Ithomia* in both appearance and behaviour. Natural selection by predators means that more of the mimic butterflies survive and leave offspring, and the *Leptalis* population gradually becomes dominated by mimics. In addition, many distasteful insects have been shown to possess eye-catching 'warning colouration' which, it is thought, registers easily in the predator's memory and reminds it not to attempt to eat an insect like that again.

Darwin encouraged Bates to publish an account of his Amazon journeys, and he put Bates in touch with his own publisher, John Murray. The following year Darwin read the proofs of Bates's book which was modestly entitled *The Naturalist on the River Amazonas*. (Bates took the river's name from the Spanish for Amazon.) The book was illustrated with lithographs by J.W. Whymper, and was an immediate success with the public when it was published in 1863.

In 1864, Bates was appointed Assistant Secretary of the Royal Geographic Society, a post which he held for the rest of his life. He contributed to several books about the fauna of South America in the years that followed, and found time to write over 50 papers and essays of his own on insects. Bates was elected President of the Entomological Society in 1869, and again in 1878. In 1881 he was made a Fellow of the Royal Society. Within a year of his death, he was decorated 'Emperor of Brazil' in recognition of the service he had paid to Brazil. Following an attack of influenza and bronchitis in 1882, Henry Bates died just one week after his sixty-seventh birthday.

Speke's journeys in East Africa, 1854–63

CHAPTER SIX
The Source of the Nile?

6.1 John Hanning Speke in 1859 from a portrait by J. Watney Wilson

By 1864, Henry Bates had earned considerable respect for his work on insects. He had succeeded Hume Greenfield as Assistant Secretary of the prestigious Royal Geographic Society and, with the President Sir Roderick Murchison, was responsible for organising the Society's meetings.

The summer meeting traditionally formed part of the British Association's annual convention, and in 1864 the Royal Geographic Society's meeting was planned for mid September, to be held in the hall of the East Wing in The Mineral Water Hospital, Bath. Bates would have had no reason to fear low attendance at this particular gathering since it promised a heated public debate on a controversial subject.

For the previous eighteen months, arguments about the opposing views of the meeting's two main speakers had both intrigued and divided Victorian society. Here at last was the chance to listen to Richard Burton and John Hanning Speke discussing their theories on the elusive source of the River Nile. This was to be the first meeting of the two men since they returned to Aden from Africa together in April 1859. The intervening four and a half years had allowed their differences of opinion to assume exaggerated importance, and for memories of the slights and insults of the past to grow into outright hatred between them.

Then, as now, the public loved a fight between famous people and so the meeting hall of the Mineral Hospital was packed with the Society's members for the preliminary meeting which was held on 15th September, the day preceding the great debate.

Burton and Speke had made two journeys together through east and central Africa. During the later one, from 1857 to 1859, they had reached Lake Tanganyika, and Speke had visited the southern shore of Lake Victoria. The expedition had been sponsored by the Royal Geographic Society, but its aims were as diverse as the personalities of the two men leading it. Burton, an experienced traveller and accomplished Arab scholar, was secretly determined to find the exact location of the source of the Nile. Speke joined the expedition to fulfil a personal ambition to go big game hunting. By the time they returned, the Nile's source had been obscured even further in controversy, and it was Speke who had become obsessed with its discovery.

Speke wasted no time in returning to England, leaving Burton convalescing from fever in Aden. He presented to Sir Roderick Murchison his opinion that the newly discovered Lake Victoria was the true source of the Nile, thus securing for himself the honour and notoriety of the discovery, and the promise of backing for a future expedition. It is hardly surprising, after the tensions and frustrations of two expeditions, that this was enough to launch the two men on their paths of rivalry and enmity.

Speke soon returned to Africa with another expedition and, early in 1863, he was able to telegraph triumphantly to the Royal Geographic Society 'The Nile is settled', believing he had undeniable evidence of the location of its source. Burton did not agree. He claimed that what little Speke had seen of Lake Victoria did not justify his conclusion that it was one huge lake. Speke's explorations of the Nile itself, he pointed out, were similarly incomplete. He had not followed the river all the way from the Ripon Falls where it left the lake, to its junction with the Blue Nile at Khartoum. Instead, Burton held the view that the Nile's true source lay further west, in Lake Tanganyika. The river which left the northern end of the lake, he claimed, was the stripling Nile. But his attempt in 1858 to reach the point on Lake Tanganyika at which the river left it had been frustrated by bad weather and his own poor health.

It was a simple and understandable disagreement. The only maps available to the two explorers were one drawn up by Ptolemy which had appeared in the second century AD, and another made by two German missionaries in 1855. Both men felt that their reputations depended upon the outcome of the Royal Geographic Society meeting in September 1864. Burton hoped to regain some of the esteem he had lost in 1858 when Speke had claimed discovery of the Nile's source, and when the Society had chosen Speke and not himself to lead the second expedition. Speke had been busy during the year since his return from Africa planning further exploration of the Nile basin and seeking support from Napoleon III. He knew that public ridicule by Burton would affect his promise of help from France, and delay the return to Africa for which he hoped.

The preliminary meeting convened in the great hall in Bath, and the two men took their places in the audience with barely a glance towards each other. As Burton's wife Isabel wrote later in her biography *The Life of Sir Richard Burton*: 'He looked at Richard, and at me, and we at him. I shall never forget his face. It was full of sorrow, of yearning, and perplexity. Then he seemed to turn to stone.'

Speke, it was observed, became increasingly restless. In the early afternoon, before the end of the meeting, he muttered 'Oh, I cannot stand this any longer', and abruptly left the hall. Less than three hours later, John Hanning Speke was dead.

He had left Bath and returned to his cousin's estate near Corsham in Wiltshire. There he had decided, as he so often did at times of overwhelming stress, to go shooting. His cousin, George Fuller, and a gamekeeper accompanied him on a partridge shoot on the estate. At about four o'clock, Fuller heard Speke's gun fire once and saw him fall to the ground. On reaching him, Fuller found Speke was badly wounded, and within fifteen minutes he was no longer alive.

Despite the controversy surrounding Speke's unexpected death, it does appear to have been a genuine accident. Speke had been standing on top of a stone wall and pulling the gun up after him, before stepping down on the other side, when one barrel had fired straight into his chest. How one so experienced in the handling and use of firearms could have neglected the cardinal rule of unloading a shot gun before climbing a wall remains a mystery, but certainly he was upset and distracted by the coming confrontation with Burton.

The next morning, when the Royal Geographic Society meeting assembled in Bath to hear the great debate, Burton and his wife waited alone on the platform. The Society's Council members emerged from a private meeting, and Murchison announced Speke's death to a stunned audience. In place of the Nile debate, Burton managed to read a lecture on Dahomey in a faltering voice. He was shocked and distraught by the loss of Speke who had been both his arch enemy and yet a special friend.

John Hanning Speke had been born on 4th May 1827 at Orleigh Court, near Bideford, a small market town in north Devon. His father William Speke, a retired army captain, belonged to an ancient Norman family. His mother was Georgina Elizabeth Hanning who came from a line of prosperous merchants. John Hanning grew up with one older brother, two younger ones and three sisters at the family's home of Jordans near Ilminster in Somerset. They were a close family and Speke's mother had a great influence upon his upbringing. His country childhood also laid the foundations for his love of natural history. His father did little to enforce education on his second son and, as a result, Speke passed through Barnstaple Grammar School and college in Blackheath, London, without distinction. According to surviving relatives interviewed by Alexander Maitland for his biography of Speke, he suffered ophthalmic attacks during childhood. Poor eyesight and a reputed preference for birds' nesting over book-learning must have further hindered his education.

In 1844, at the age of 17, Speke joined the 46th Bengal Native Infantry in India. He served bravely under Lord Gough in the Punjab War, gaining medals for four separate battles, and later he fought in the Sikh War. Peacetime duties allowed Speke to relieve his boredom of camp life by travelling extensively in northern India. These journeys also enabled him to shoot and collect the animals and birds of the region.

It was during these years that Speke developed his own particular talents as a naturalist. Then, and throughout his life, he was first and foremost a sportsman, enthusiastically pursuing every opportunity for shooting. But, unlike many of his contemporaries, he also took a keen interest in the game, and learnt about the habits and behaviour of the animals he hunted. He kept the less damaged specimens of the birds and animals he shot, and learned how to skin and preserve them himself. Quite often, as we see even in modern naturalists, someone who is absorbed by the animals he shoots becomes a first class biologist and conservationist. But Speke seems to have lacked whatever it is that is needed to complete this transition. He showed no signs of feeling sufficiently inquisitive to resort to books for more information about his specimens. Perhaps it was his upbringing, which had caused 'the young man's thinking... to develop along emotional rather that rational lines', and his neglected education that denied Speke the chance ever to obtain anything approaching an academic interest in zoology. He shot primarily because he enjoyed the sport, and he collected specimens more as trophies than with the curiosity of a scientist. And yet, as we shall see, when his specimens were passed on to professionals for examination and identification, they made valuable contributions to zoological knowledge.

Aided by the approval of his Commander-in-Chief in India, Sir William Gomm, Speke was allowed generous periods of leave. He travelled almost alone and without luxuries, often quite literally living off the land as the locals did, and found he could stretch his pay and prolong the journeys. He wandered high into the Himalayas and through little-known parts of Tibet, shooting and mapping as he went. His specimens, he decided, would form the beginning of a museum when he returned to Jordans, and, as Maitland puts it, 'there is no record, so far as is known, of [Speke's parents] having objected to their son's proposal'! Presumably they agreed as in 1864 Speke was able to write: 'There are now but few animals to be found in either India, Tibet, or the Himalaya Mountains, specimens of which have not fallen victims to my gun. Of this the paternal hall is an existing testimony.'

Memories of the special contentment of these days stayed with him all his life, as the following passage written in 1864 shows:

> Without exception, and after having now shot over three quarters of the globe, I can safely say, there does not exist any place in the whole wide world which affords such a diversity of sport, such interesting animals, or such enchanting scenery, as well as pleasant climate and temperature, as these various countries of my first experiences; but the more especially interesting was Tibet to me, from the fact that I was the first man who penetrated into many of its remotest parts, and discovered many of its numerous animals.

By 1849, Speke had reached the half way point in his army service, and his thoughts turned to his three years of leave due in 1854. He became increasingly fascinated by the idea of shooting and exploring in central Africa. He later wrote:

> My plan was made with a view to strike the Nile at its head, and then to sail down that river to Egypt. It was conceived, however, not for geographical interest, so much as for a view I had in my mind of collecting the fauna of those regions.

He makes no particular mention, here or anywhere else, of an ambition to find the source of the Nile, and the mystery of its whereabouts may not have impressed him until he met Burton.

He managed to continue enlarging his collection of Indian specimens, and to save some of his pay. At the start of his leave, on 4th September 1854, he left Calcutta for Aden, having given, it seems, very little thought as to how he was actually to reach Africa, or survive once he got there. However, he had had the foresight to provide himself with some cheap guns, revolvers, swords, cutlery, beads and cottons to barter with 'the simple-minded negroes of Africa'.

On reaching Aden, Speke went directly to the newly arrived political Resident, Colonel James Outram, and outlined his plans with great enthusiasm. He proposed, he said, to travel to Somaliland and then south to a relatively unknown area which had been marked on Ptolemy's map as *Lunae Montes* or Mountains of the Moon. In this region, which Speke presumed to be not unlike the Himalayas, he planned to collect specimens and carry out mapping surveys. Despite Speke's considerable charm, Outram insisted that travel among the warring Somaliland tribes would be nothing but suicidal, and he refused Speke the necessary letters of introduction. Finally, hoping to relieve himself of the persistent requests, he sent Speke to meet Lieutenant Richard Burton of the 18th Bombay North Infantry who had arrived in Aden a few months earlier.

In his introduction to Burton's *The Lake Regions of Central Africa*, Alan Moorhead wrote:

> Burton was a hopeless case. He was one of those men in whom nature runs riot: she endows him with not one or two but twenty different talents, all of them far beyond average, and then withholds the one ingredient that might have brought them to perfection — a sense of balance and direction.

Educated in France and Italy, and at Oxford,

C1.

C2.

C4.

C5.

C6.

C7.

C9.

C8.

C10.

C11.

C12.

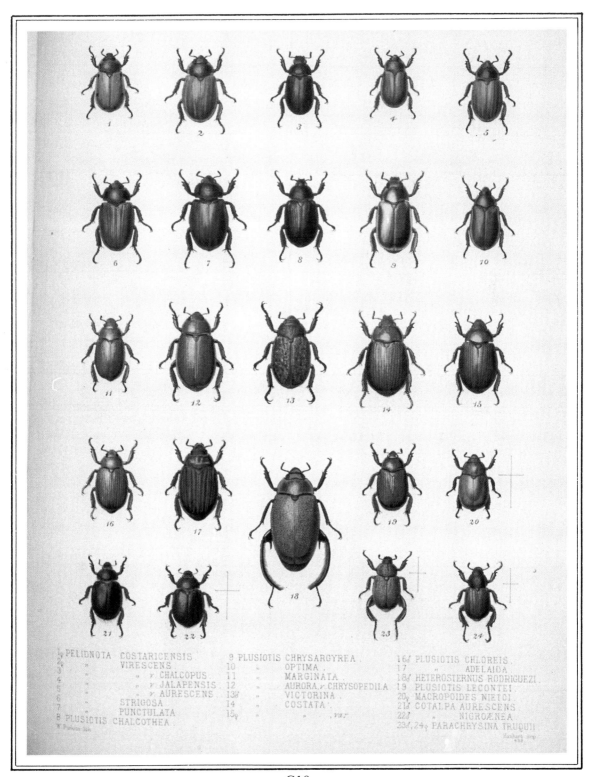

1. PELIDNOTA COSTARICENSIS.
2. " VIRESCENS
3. " "
4. " v. CHALCOPUS
5. " v. JALAPENSIS
6. " v. AURESCENS
7. " STRIGOSA
7. " PUNCTULATA
8. PLUSIOTIS CHALCOTHEA.
9. PLUSIOTIS CHRYSARGYREA.
10. " OPTIMA.
11. " MARGINATA
12. " AURORA, v. CHRYSOPEDILA
13♂. " VICTORINA.
14. " COSTATA.
15♀. " " var.
16♂. PLUSIOTIS CHLOREIS.
17. " ADELAIDA
18♂. HETEROSTERNUS RODRIGUEZI
19. PLUSIOTIS LECONTEI
20♀. MACROPOIDES NIETOI.
21♂. COTALPA AURESCENS
22♂. " NIGROÆNEA
23♂, 24♀. PARACHRYSINA TRUQUII

C14.

C15.

C17.

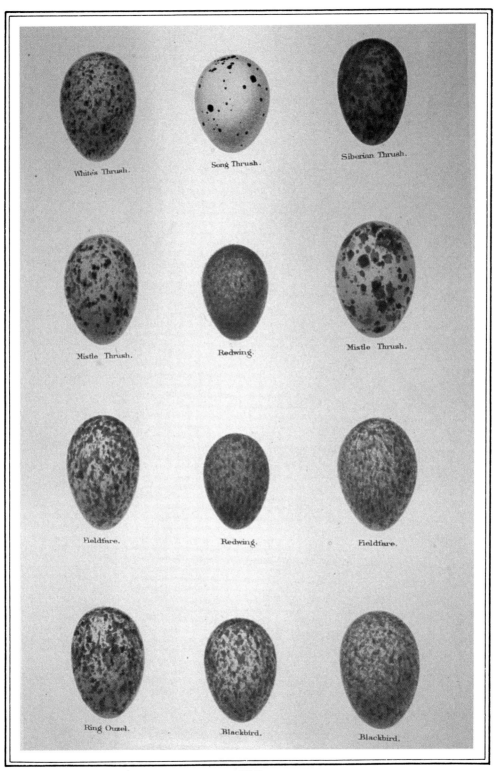

C16.

C18.

C19.

C20.

C21.

Burton had served seven years in the Indian Army. None of the many books about him have been able to fathom his character or his fierce need for excitement in his life. He was an exceptional linguist and an Arab scholar and, being acutely observant, he had made himself an expert on the customs, geography, botany, geology and meteorology of the countries through which he had travelled. Moorhead describes Burton as 'an intensely fastidious and scholarly man ... No other explorer had such a breadth of reference, or had read so much or could write so well.'

By the time he arrived in Aden in 1854, at the age of 34, Burton was already famous for his visit to Mecca, in today's Saudi Arabia. By assuming and successfully carrying off an elaborate Arab disguise, he had become the first European to enter the religious city, forbidden to non-Moslems.

Burton had obtained the backing of the Royal Geographic Society for a visit to Africa. His aim, or so he had claimed, was to open up a potential trading route from Somaliland or northern Ethiopia southeast to Zanzibar and thence westwards to the Atlantic. His private intention was to explore the lakes and mountains of central Africa described by a missionary, Johann Krapf, who had visited the region in 1849. Burton had met Krapf in Cairo in 1853 and was convinced that the Nile flowed from one of the lakes. What better challenge did he need than to clear up, once and for all, the mystery surrounding the exact source of the Nile? So he had travelled to Aden, en route to Africa, with two friends, also veterans of the Indian Army, Lieutenant George Herne and Lieutenant William Stroyan.

Had it not been for the death of the expedition's fourth member, which occurred in England a few months earlier, Burton might never have agreed to take Speke along with him. The two men were attracted by mutual admiration when they first met although Burton was immediately scornful of Speke's naivety in matters of African travel, and his lack of linguistic or scientific knowledge. He judged Speke's character shrewdly:

> To a peculiarly quiet and modest aspect — aided by blue eyes and blond hair — to a gentleness of demeanour, and an almost childlike simplicity of manner which at once attracted attention, he united an immense fund of self-esteem, so carefully concealed, however, that none but his intimates suspected its existence.

He adds a description of Speke's almost Scandinavian appearance and tall wiry body which suggested he was capable of untiring endurance.

Burton insisted that the expedition members should travel in disguise to avoid attracting the attention of Somali tribesmen. Reluctantly, Speke bought an uncomfortable Oriental costume including a long, close-fitting gown and heavy turban. He also equipped himself with various foods and materials as presents for the natives and, for his own use:

> rifles, guns, muskets, pistols, sabres, ammunition in great quantity, large commodious camel-boxes for carrying specimens of natural history, one sextant and artificial horizon, three boiling-point and common atmospheric thermometers, and one primitive kind of camera obscura ... made in Aden under the ingenious supervision of Herne.

It was agreed that Speke should explore the Wadi (or Wady) Nogal region at the extreme northeastern tip of Somaliland. Lieutenant Herne would collect baggage mules in Berbera, whilst Stroyan was sent off to survey parts of the nearby coast. The three men were to rendezvous at the annual fair held at Berbera in mid January 1855. Meanwhile, Burton, heavily disguised as usual, was to travel west to Harar, the religious capital of Ethiopia and another challenging 'forbidden city'.

Speke's first African journey is described in his book *What Led to the Discovery of the Source of the Nile*, and his diary appeared in Burton's own account of the expedition, *First Footsteps in*

East Africa. His five-month journey was disappointingly unsuccessful. Afterwards Speke never quite got over the feeling that he had been tested by Burton and had failed. His expedition was beset from the start by language barriers, the disloyalty and disobedience of the native porters and guides, and by many other difficulties. Speke neither reached the Wadi Nogal to collect the red soil samples alleged to contain gold dust, nor did he manage to find any baggage animals to buy for Burton's expedition. But he was able 'to make a valuable collection of the Fauna' which was forwarded later to the Curator of the Royal Asiatic Society's Museum, Calcutta.

The geography, scenery, meteorology and local customs of the region are thoroughly, if rather unimaginatively, described in his diary, and of course he gives special atttention to the game. Most of his time was spent in the area of Las Koreh. The Nogal valley had been described by Lieutenant Crutenden of the Indian Navy, who had surveyed the coast in 1848, as a 'very fertile and beautiful valley abounding in game. ... Water abundant. Elephants numerous'. But it still lay 120 miles to the south.

On the coastal plain, Speke found many antelopes and birds. While being delayed by a local chieftain one day, he shot a small gazelle which he suspected had never been seen before. The species was later named *Gazella spekei* or Speke's gazelle, and even today is rare and little known. Speke's gazelles live in small herds of never more than 20 animals, and are now confined to the arid plateau of central Somalia, although Speke's specimen clearly was shot on the coastal plain.

When he pushed inland across the 6000-foot mountain range, Speke found fossil shells in the limestone rockface and saw hyenas, jackals, more antelopes, hares and rats but, surprisingly perhaps, he collected only one type of snake. On 29th November he wrote:

> The Somali brought a leopard into the camp, which they said they had destroyed in a cave by beating it to death with sticks and stones. They have a mortal antipathy to these animals, as they sometimes kill defenceless men, and are very destructive to their flocks.

He recorded 'Saltiana antelope', or 'Salt's antelope' as he also calls them. These were probably Salt's or Phillip's dik-diks which still live in the bush and scrub habitat of the Somali mountainsides. Speke admired the local hunters who tracked the big game, including rhinoceros, with only bows and spears. Rhinoceroses are no longer found in Somalia. Even in Speke's time, their hides were prized for making shields and they were said to be becoming scarce.

Once he reached the plateau of the interior, Speke overcame his disappointment at finding it to be a desert-like plain, and spent a few days collecting. He saw great numbers of antelopes, gazelles and birds including large troops of ostriches, bustards, florikan sandgrouse and partridges. He shot lynx, and more dik-diks which the Somalis could chase and catch on foot. He heard various tales of local hunting methods including the two ways to catch an ostrich. Either, Speke records, one throws down 'some tempting herb of strong poisonous properties, which they eagerly eat and die from'. Otherwise, armed with no more than a Somali pony, 'wonderfully hardy and enduring but not swift', and three days' supplies, the hunter follows the ostriches staying close enough to keep them in sight and on the move, but never bolting them. Then 'at night the birds stop in consequence of the darkness, but cannot feed.' Amazingly, it is recommended that the hunter courteously 'dismounts to rest and feed with his pony, and resumes the chase the following day'. By the third day, with the pony and man still fresh but the ostriches weak 'from constant fasting', the hunter is able to ride up to them and knock them down!

While he was in the mountains, Speke collected a small lizard which turned out to be another new species. Although, to Speke's dismay on his return, Burton sent all his specimens to Edward Blyth, Curator of the Asiatic Society's Museum in Calcutta, he insisted on the lizard being named *Tiloqua burtoni*, after Burton. The collection included 20 specimens of mammals, 36 birds, three reptiles, one fish, one scorpion and three beetles. He had collected another new reptile in addition to the lizard, a snake named *Psamophis sibilans*.

Not all the strange species were new ones. A rat 'with a bushy tail, ... plumper [than a squirrel] in the face and body, like a recently born rabbit' which he shot near the coast was thought to be a new species, and named by Blyth *Pectinator spekei*. However, this was later found to be a member of the cane rat group of the genus *Thryonomys* which is widely distributed throughout Africa and has complicated taxonomy. Several mysteries exist in Speke's records and one surrounds an all black hyena-like animal with a white tail. This, he observed, was slightly smaller than the spotted hyena found in Somaliland, but shaped more like a wolf. First he saw a single animal and then three hunting together near the coast, but on neither occasion was he able to shoot a specimen so that to this day the species' identity remains unknown.

Despite the frustrating delays and confrontations with the expedition's guides and porters, Speke developed a great respect for the Africans. He admired their appearance, and their bravery and skill in hunting. In this opinion he was to contrast with Burton who admired only the Arabs in Africa, and could never fully appreciate the native tribesmen.

Speke returned to the coast, and set sail for Aden on 19th February 1855. There he met Burton who had successfully reached Harar and spent three days in the city. Six weeks later the expedition members were reunited at Berbera on the Somaliland coast where the annual fair was now drawing to a close.

Unknown to the four men, rumours were circulating among the natives that the British were about to recommend putting an end to the slave trade, upon which the local Arabs thrived. In addition, news of the trouble caused to Speke's expedition by his deceitful guide Summunter, who had been tried and heavily fined in Aden upon his return, did nothing to improve the foreigners' popularity. Unfortunately, when the large caravan in whose safety they had planned to travel left Berbera, the expedition chose to stay behind and wait for the arrival of a boat from Aden bringing some letters and equipment from England. The silence and emptiness of the deserted fair ground was unnerving, so they welcomed the arrival of a small Arab vessel whose captain had come to see what was left of the fair. What happened next can only be told in Speke's own words:

> At the usual hour we all turned in to sleep, and silence reigned throughout the camp. A little after midnight, probably at one or two, there suddenly arose a furious noise, as though the world were coming to an end: there was a terrible rush and hurry, then came sticks and stones, flying as thick as hail, followed by a rapid discharge of firearms, and my tent shook as if it would come down.

Soon Burton and Speke were fighting for their lives side by side in the ruins of the former's tent.

> In another instant I was on the ground with a dozen Somali on top of me. The man I had endeavoured to shoot wrenched the pistol out of my hand, and the way the scoundrel handled me sent a creeping shudder all over me. ... I feared that they belonged to a tribe called Eesa, who are notorious ... for the unmanly mutations they delight in. Indescribable was my relief when I found that ... the men were in reality feeling whether, after an Arab fashion, I was carrying a dagger between my legs.

Speke was taken aside, still tied up, and given water to drink while the Somalis looted and destroyed the camp. After dawn, some tribesmen came up to Speke and he was speared through the shoulder, the hand and the thigh, the last of these blows being so violent it pinned him temporarily to the ground.

With the action of lightning, seeing that death was inevitable if I remained there a moment longer, I sprang upon my legs ... and ran over the shingly beach towards the sea like wildfire.

Speke managed to evade the half-hearted pursuit and later, exhausted and weak from loss of blood, he was brought onto the Arab vessel, which fortunately was still moored in the harbour. Here he found the other expedition members. Herne was almost unhurt, Burton had received a spear thrust through his mouth but Stroyan, everyone's favourite, had been killed. The crew were sent to the deserted campsite to retrieve what was left of the luggage and equipment, and to bring back Stroyan's body. Then the expedition returned sadly and painfully to Aden.

Speke's wounds healed amazingly fast and little over a month later he was able to return to England. The Crimean War had started the previous year, and both Speke and Burton served in it. However, Speke appears to have continued planning to return to Africa, this time with the specific intention of searching for the Nile's source. Soon after the war ended in 1856, Burton invited Speke to join him in another expedition to central Africa, and Speke immediately cancelled plans to travel through the Caucasus Mountains shooting game, and accepted.

By December, the two men were in Zanzibar backed with a grant of £1000 from the Royal Geographic Society. However, relations between them had not improved. Speke was still feeling humiliated by his failure to reach Wadi Nogal, and was furious with Burton for sending the specimens he had collected in Africa to Blyth in Calcutta. Burton had angered Speke further by publishing his diary, translated into the third person, as an appendix to his *First Footsteps in East Africa* and by implying, in the same book, that Speke had turned to run away from the attacking Somali tribesmen at Berbera.

After a preliminary trip up the Pangani River, where Speke had his 'first flirtations with the hippopotami', and a trek into the nearby hills taken, to Speke's disgust, at so fast a pace he was unable to shoot anything, the expedition returned to Zanzibar. They set out again for the mainland, and started on their journey inland on 27th June 1857. Speke's diary, published later in his book *What Led to the Discovery of the Source of the Nile*, gives relatively few details of the animals he saw and collected on this trip. It may have been because Burton disapproved of what he regarded as time being wasted on hunting so that Speke was not able to do as much shooting as he would have wished. He also suffered during the expedition from fevers, eye trouble and other ill health, as indeed did Burton.

Burton's two volume account of the expedition, *The Lake Regions of Central Africa*, is a mine of information about the geography, peoples, customs and economy of east and central Africa. The style is detailed yet readable although Speke is referred to throughout merely as 'my companion'. Here is how Burton saw the savannah through which they passed after leaving the coastal plain:

Long lines, one bluer than the other, broken by castellated crags and towers of most picturesque form, girdled the far horizon; the nearer heights were of a purplish-brown, and snowy mists hung like glaciers about their folds. The plain was a park in autumn, burnt tawny by the sun ... Here the dove cooed loudly, and the guinea-fowl rang its wild cry, whilst the peewit chattered in the open stubble, and a little martin, the prettiest of its kind, contrasted by its nimble dartings along the ground with the condor wheeling slowly through the upper air. The most graceful of animals, the zebra and antelope, browsed in the distance: now they stood to gaze upon the long line of porters, then, after leisurely pacing, ... they bounded in ricochets over the plain.

No wonder Burton detested Speke and his gunshots disturbing such a tranquil landscape. Later Burton lists the fauna of the area, with a keen eye for detail.

It took them four and a half months to cover 600 miles to Kazeh, now the city of Tabora in central Tanzania. Both men were weak from bouts of fever and lack of food. Their supplies had run low, and the savannah game had been scarce and difficult for even Speke to shoot. They were able to recuperate at the house of a friendly Arab in the town, Sheikh Snay, who was a trader in slaves and ivory. He told them that lying to the west was not one lake, as they had assumed from the map drawn up by the missionaries Rebmann and Ebhardt which they had seen hanging in the Royal Geographic Society in London; instead, Snay assured them, there were three lakes. Due west lay the sea of Ujiji or Lake Tanganyika. South of this was Lake Nyassa and to the north a separate, much larger lake called Ukerewe which Speke immediately guessed was the source of the Nile. When Snay confirmed that a river ran out of Ukerewe, Speke became all the more impatient to explore it.

However, when they set out once more on 14th December, they headed west on Burton's insistence, and not north. Both men became ill again, Burton with a severe attack of malaria and Speke with ophthalmia which he partly remedied by wearing 'stained-glass spectacles'. The 250 mile journey to Ujiji on the shores of Lake Tanganyika took them about two months. Although Burton arrived there weak and sick he was impressed by the beauty of the lake and wrote:

> In front stretch the waters, an expanse of the lightest and softest blue . . . and sprinkled by the crisp east wind with tiny crescents of snowy foam. The background in front is a high and broken wall of steel-coloured mountains, here flecked and capped with pearly mist, there standing sharply pencilled against the azure air.

Again he gives a detailed list of the animals and birds in the area right down to the insects, leeches and millepedes.

While Burton recovered his strength, Speke crossed over to the island of Kasenge near the west shore of the lake to try and hire a dhow from Sheikh Suleyim. His health had improved enough for him also to appreciate the lake:

> Here tall aquatic reeds diversify the surface, and we are well tenanted by the crocodile and hippopotami, the latter of which keep staring, grunting, and snorting as though much vexed at our intrusion on their former peace and privacy.

His sight was sufficiently good for him to be able to collect some interesting shells on the lake shore, although he was still unable to shoot accurately.

> Here a herd of wild buffaloes, horned like the Cape ones, were seen by the men, and caused some diversion: for, though too blind myself to see the brutes at the distance that the others did, I loaded and gave them chase . . . and sprang some antelopes, but could not get a shot.

To make matters worse, while camping on the island of Kivira, a small beetle entered Speke's ear and dug itself through his ear drum and into the middle ear. The intense irritation that this caused him drew the inflammation away from his eyes, but he was forced, when successive applications of oil, salt and melted butter had failed to extract the buzzing insect, to lance his ear with the tip of a penknife. This killed the beetle but the resulting suppuration extended down his face and neck to his shoulder. The tumour that developed inside his ear left him slightly deaf for the rest of his life.

Sheikh Suleyim told Speke of a river which, he said, left the lake at its northern end. Speke returned to Burton with this news, but having been unable to hire any boats. By late April 1858, travelling together once more, they had reached Uvira on northern Lake Tanganyika in two large canoes hired at an extortionate rate by Burton. A lack of stores and Burton's poor health forced the expedition to return to Kazeh and the hospitality of Sheikh Snay without reaching the northern river which,

6.2 On one occasion, Speke was charged by buffaloes three times in one day

Burton had become convinced, was the beginning of the Nile.

Speke held fast to his opposing view. In June 1858 he left Burton in Sheikh Snay's company, in which Speke always felt out of place, and headed north towards Lake Ukerewe with a small expedition of his own. In this parting of the ways, as Speke's biographer puts it, 'Burton finally defined his role as a traveller and ethnologist, while Speke, by responding to the challenge of the great unknown, confirmed his as an explorer'.

His excitement made Speke force the pace of the march and gave him little time for game. Almost his only comment is a vague 'Here I saw a herd of hartebeests, giraffes, and other animals, giving to the scene a truly African character.'

By the beginning of August the expedition was following a creek leading into the lake.

> Would that my eyes had been strong enough to dwell, unshaded, upon such scenery! but my French grey spectacles so excited the crowds of sable gentry who followed the caravan, and they were so boisterously rude, stooping and peering underneath my wide-awake to gain a better sight of my double eyes, as they chose to term them, that it became impossible to wear them.

Instead, poor Speke approached what he expected to be the culmination of four years' ambition riding on a donkey, with the spectacles in his pocket and his eyes closed against the glare of the sun.

However, when they first sighted the lake itself, on the early morning of 3rd August, he was able to appreciate the view with triumph. Speke renamed the lake Victoria N'yanza and stayed just a few days surveying and mapping the area. Then he hurried back to Kazeh to tell Burton of his discovery of this huge new lake. Despite the furious pace of the expedition, he noted more details of game on his journey than at any other time on the expedition. He includes a description of hunting hippopotamuses which ended up with Speke flat on his back in the bottom of the canoe, one of his guns in the water and the native with him being thrown briefly onto the angry hippo's back.

Burton was sceptical of Speke's find and, in an 'atmosphere of strained cordiality and mutual distrust', the two men left Kazeh for Zanzibar in early September. An argument en route about the expedition finances did nothing to bridge the gap between them. But later, when Speke became deliriously ill with a fever, he was nursed by Burton. As we have seen, Speke travelled home without delay from Aden and brought news of his discovery of Lake Victoria to Sir Roderick Murchison before Burton had even arrived back in England.

He was invited to speak at the Royal Geographic Society meeting on 13th June 1859. James McQueen, a respected geographer and staunch friend of Burton's, questioned Speke after his talk and cast the first public doubt on his theory that Lake Victoria was the source of the Nile. Speke's views immediately became the subject of controversy and speculation especially when Burton returned home with such contrasting opinions. It was Burton, without Speke, who received the Society's prestigious Gold Medal later that year, and this made Speke even more determined that the next expedition should prove him right and Burton wrong about the source of the Nile.

When Speke returned to Africa in 1860, he chose James Grant to accompany him. He had met Grant briefly in India during 1847 when Grant had been a captain in the Bengal Army. Speke regarded him as a good soldier and, even more important, as a good sportsman. Maitland describes Grant as a cautious, patient and shrewd Scotsman, and a highly talented artist with a flair for botany. The difference in characters was sufficient for Speke and Grant to establish a strong and long lasting friendship.

The two men arrived at Zanzibar on 17th August, and on 2nd October they started out for Kazeh with a large mixed caravan. This expedition differed from the previous one as Speke was intent on collecting specimens and allowed himself plenty of time for shooting. He had arranged for his collections to be returned periodically to P. L. Sclater at the Zoological Society of London. There they would be examined, identified and stored until his return. He had sent off the first case of specimens, collected in Zanzibar, to Sclater with an accompanying letter saying:

> In furnishing these specimens, I am sorry that I cannot give you more particulars about them, since the necessities incidental to the organisation of the expedition have occupied my time too much for me to make the collection with my own hands. I have, therefore, employed my Hottentot guards both in shooting and in skinning them.

Unfortunately, the glass bottles in one of the cases had smashed by the time it reached London, the preserving spirit had evaporated and the specimens were ruined.

They settled down to an expedition routine. As Speke records in his *Journal of the Discovery of the Source of the Nile*:

> My first occupation was to map the country. This was done by timing the rate of march with a watch, [and] taking compass bearings.... On arrival in camp every day came the ascertaining, by boiling a thermometer, of the altitude of the station above sea level; of the latitude of the station by the meridian altitude of a star taken with a sextant; and of the compass variation by azimuth.

These measurements, crude by today's navigational standards, were vital to the aims of the expedition and for preventing them becoming completely lost. Meanwhile Grant made temperature measurements, a botanical collection and water colour sketches which later formed the basis for the illustrations in Speke's *Journal*.

Speke found particularly rewarding hunting grounds in the foothills of the coastal range. By the end of October he was able to return more specimens to Sclater in London. These included the heads of an antelope and a warthog, the skins of several monkeys and birds, and some fish. Later Speke reluctantly decided to part with the camera they had brought with them, saying in his diary: 'Had I allowed my companion to keep working it, the heat he was subjected to in the little tent while preparing and fixing his plates would very soon have killed him.'

The land was full of animals and birds, and hippopotamuses 'snorting as if they invited an attack'. He shot at an unusual striped eland and in the same spot saw some other animals the size of hartebeest 'only cream-coloured, with a conspicuous black spot in the centre of each flank' with which he was unfamiliar. A specimen of the eland, which Speke succeeded in collecting later, was pronounced by Sclater to be the first one of its kind to reach Europe although both Livingstone and Kirk had seen it in Africa. It was duly named *Oreas livingstonii* (later changed to *Taurotragus oryx*) or Livingstone's eland. The other animals were almost certainly topi which are migratory and gather into herds of many thousands in the dry season.

Speke also found time for some hair-raising rhinoceros shooting by moonlight. At one moment he turned round to the boy who was loading guns for him only to find his nerve had given and he had fled up a tree. He collected another new species, this time a kind of gazelle with beautifully curved horns forming a heart shape from the front view. This was named *Gazella granti* after Grant.

Speke and Grant arrived in Kazeh late in January 1861 and remained there for nearly two months. Another batch of specimens were sent off to the Zoological Society of London with an apologetic note from Speke:

> It is next to impossible to stuff and take care of animal-specimens properly when travelling with a large caravan, destined for a long journey, and in constant motion. You must therefore take [the specimens] as you find them, for the present; but I hope they will interest you sufficiently to direct your attention more particularly to these regions; for I am convinced in my mind that the great varieties of animal life, large and small, which are to be found here would fully repay any trouble or expense in procuring and the Society would do well if they could find competent men who would voluntarily spend a few years collecting them.

Elsewhere in the letter he almost reveals a sense of humour when he describes the noise made by an excited zebra as 'something like a sheep trying to bleat with a bad cold and cough'!

Speke had made a special trip to try and collect two more unusual antelopes. These

6.3 While Speke was staying in Karagwe, the king presented him with a live sitatunga or waterbuck, a species of antelope previously unknown

were the blanc bok and the large sable antelope which has huge ridged horns curving over its back. The latter, Speke told Sclater, were 'very scarce, for in six days' constant shooting I only saw one.' The blanc bok was evidently also very rare; Speke mentions it only at Kazeh. He obtained just one specimen, and this he had to retrieve from a lion because he was unable to shoot one. From Speke's description, it is possible that this species was the bluebuck or *blaauwbok* which is related to the sable and roan antelopes of the genus *Hippotragus*. The bluebuck is known mainly from further south in Africa where it used to have a restricted distribution and small population. However, the species was particularly vulnerable to large scale shooting, and is thought to have become extinct about the beginning of the nineteenth century. Unfortunately, more valuable information was lost when upon returning to Kazeh from a hunting trip, Speke found a large part of his collection of birds had been destroyed by insects and had to be thrown away.

In November the expedition finally reached the kingdom of Karagwe, to the east of Lake Victoria, in today's northern Tanzania. They had been delayed by illness and by the chiefs of tribes along the route who had refused to give them men. They remained as guests of King Rumanika for nearly two months, and during that time Speke filled his diary with notes on the extraordinary life of the palace. Obesity, for example, was considered the height of beauty and fashion. Speke's description of taking measurements of the seal-shaped princess raised quite a few Victorian eyebrows when it appeared in his *Journal*.

King Rumanika presented Speke with a live specimen of a sitatunga or waterbuck. This species, previously unknown, lives only deep inside swamps or papyrus marshes. It is a truly aquatic antelope and, as Speke noticed when he first saw one, it has developed deeply forked cloven hooves which can be splayed out to help support the animal on soft ground. The sitatunga is shy and mainly nocturnal, though Speke reports its skin was highly prized by the Africans. Sclater described the new species from skins and heads he received from Speke, and named it *Trangelaphus spekii*.

When Speke left Karagwe in January 1862, Grant remained behind with a fever. Speke pressed on northwards, although always out of

6.4 *Speke presents his 'spoils' to King Rumanika, the heads of three white rhinoceroses*

sight of the lake. He crossed rolling grassland with the vegetation often towering above his head, and passed through marshy valleys swarming with wildfowl. He reached the Kitangule River on 16th January, and headed round the north of Lake Victoria to the palace of the next tribal king, Mtesa of Buganda. At one camp site he records: 'At night a hyena came into my hut, and carried off one of my goats that was tied to a log between two of my sleeping men.'

Grant caught up with Speke at Mtesa's palace, close to the site of the modern city of Kampala. The expedition stayed nearly five months with the king and his queen, whom Speke describes as eating 'with great relish, making a noise of satisfaction like a happy guinea pig.'

When they finally extricated themselves from Mtesa's alternating threats and friendship which had delayed them in Buganda, Speke and Grant moved along the north shore of Lake Victoria towards the point at which the Nile flowed from it. Speke shot two zebras for Mtesa who, by Ugandan law, was the only one allowed to keep 'their royal skins'. These were Burchell's zebras but the expedition collected a previously unknown subspecies, later named Grant's zebra. Speke also collected a leucotis or n'sunnu antelope, the first one he had seen although they later became very common. This was a kob, probably Thomas' kob, now one of the commonest antelopes of the savannah.

As they neared the Nile, Speke rather surprisingly sent Grant northwards to make contact with the next local king, while he pushed on alone to the river. On 21st July 1862 he wrote in his diary:

> Here at last I stood on the brink of the Nile. Most beautiful was the scene; nothing could surpass it! It was the very perfection of the kind of effect aimed at in a highly-kept park; with a magnificent stream from 600 to 700 yards wide, dotted with islets and rocks, the former occupied by fishermen's huts, the latter by sterns [terns] and crocodiles basking in the sun, flowing between fine high grassy banks, with rich trees and plantains in the background, where herds of n'sunnu and hartebeest could be seen grazing, while the hippopotami snorting in the water, and florikan and Guinea-fowl rising at our feet.

That evening he 'strolled in the antelope parks, enjoying the scenery and sport excessively', a well deserved peaceful interlude in the struggles of his incredible journey. On his way back to the village he shot a large goatsucker, with inner primaries (central wing feathers) twice the length of its body which streamed out behind it as it flew. This was yet another species to be named initially after Speke. Later it was realised that the species, now known as the pennant-winged nightjar, had been described and named by John Gould in 1838 from a specimen collected in Sierra Leone.

The expedition struggled on through a jungle of tall grass beside the river and arrived, a week later, at a splendid set of rapids and waterfalls which Speke christened the Ripon Falls after Lord Ripon, Murchison's successor as President of the Royal Geographic Society. Speke spent a day fly fishing in the river and 'felt as if I only wanted a wife and family, garden and yacht, rifle and rod, to make me happy here for life, so charming was the place!' Later a carefully worded plaque was placed at the Ripon Falls in memory of Speke:

> Speke discovered this source of the Nile on 28th July 1862.

But, as Moorhead points out, the implications of the inscription no longer matter: 'The Ripon Falls have now been submerged beneath a hydro-electric dam, and somewhere in the green depths of the great river the place where Speke's plaque used to stand has been obliterated for ever.'

He shot his first n'samma antelope near the river; this was very like the common waterbuck (or ellipsiprymnus as Speke calls it), but lacked the distinctive white band across the rump. Two heads and a foot of this species were sent

6.5 Speke discovered this goatsucker, a species of nightjar, on the north shores of Lake Victoria

to Sclater in London, and he identified them as belonging to the defassa waterbuck. For a while Speke and his men travelled by boat on the Nile before striking away from the river for the kingdom of Unyoro. Here he was reunited with Grant and there followed a far from comfortable stay of two months in the palace of King Kamrasi. During this time, Speke continued collecting and also 'purchased a small kitten, *Felis serval* (a serval), from an Unyoro man, who requested me to give it back to him to eat if it was likely to die, for it is considered very good food in Unyoro.' Finally, they were able to bribe Kamrasi into allowing them to continue on the final stage of their journey.

When Speke and Grant had left England, two and a half years earlier, they had arranged with John Petherick, a friend and fellow African explorer, that he should travel up the Nile from Khartoum and meet the expedition at the mission of Gondokoro in southern Sudan. As it turned out, the Royal Geographic Society was little help with financing the expedition. Most of the expenses of the long journey made by Petherick and his dauntless wife had to be covered from their own pocket, supplemented by the proceeds of ivory trading. The Pethericks waited patiently for the long-overdue explorers. When Speke and Grant did finally arrive at Gondokoro, in mid February 1863, the Pethericks were away and instead they met Samuel Baker, an English sportsman and ivory trader. Speke became convinced that Petherick had let him down by not being there to meet him, and refused all offers of help when Petherick returned to the village. Sadly, Speke's unjustified hostility was enough to ruin the Pethericks' reputation for life.

Grant and Speke sailed down the Nile, using Baker's and not Petherick's boats, to

Khartoum. There Speke found a letter from the Royal Geographic Society informing him he had been awarded their Gold Medal in 1861. In return he sent his famous telegram saying 'The Nile is settled'. That summer the two explorers returned to England, to be welcomed as heroes. The sad ending to the story, with its misunderstandings followed by Speke's premature death, has already been told.

The scientific results of the expedition were most rewarding. An appendix to Speke's *Journal* lists the 23 species of large game shot between mid September 1860 and late December 1862, a total of 75 animals. The commonest species, or probably the most easy to shoot, were zebra (11), white and black rhinoceros (6 each), Buffalo, Grant's gazelle, defassa waterbuck and pallah bok (4 each), and Salt's dik-dik, blue (?) duiker and bush buck (3 each). Sclater, reporting to the Zoological Society on Speke's East African mammal collection, mentions 39 species, including 16 antelopes, and gives descriptions of the sitatunga and Livingstone's eland. This impressive collection of specimens was never enlarged because of Speke's tragic death. It is interesting to speculate on what geographical and zoological finds he might have made had he returned to Africa as he had planned.

Grant's modest botanical collection also proved interesting. According to Speke: 'This unique collection is the first that was ever made *by the drying process* in the interior west of Zanzibar.' Before this flowers and plants usually had been preserved by being crushed in heavy and unwieldy presses which would have been unsuitable for expedition use. Instead, Grant first dried his specimens, and then stored them neatly between protective sheets. Dr T. Thomson examined Grant's collection at the Botanical Gardens at Kew on his return to London, where it was also 'highly commended' by Sir Joseph Hooker, Kew's Director. There were 750 species of plants 'carefully ticketed, with numbers attached referring to a note-book in which all essential points of habit and uses are entered.' These had been collected from Zanzibar to southern Egypt, and about 80 of them were previously unknown. The new species were mostly members of existing African genera but two were in genera known before only from India, and there were several plants belonging to completely new genera.

Speke's reputation received a battering during the controversy surrounding his views which continued long after his death. Burton published his own theories in *The Nile Basin* (1864). The book's co-author, James McQueen, was one of Speke's most outspoken critics, and he took the opportunity to launch an unnecessarily vehement attack on Speke's conduct, accompanying it with exaggerated criticism of the expedition's discoveries.

Extraordinary as it may seem, Speke was not vindicated until 11 years after his death. In 1871, David Livingstone and Henry Stanley explored Lake Tanganyika. Stanley returned to Africa four years later with a large expedition lavishly equipped and funded by the London *Daily Telegraph* and *The New York Herald*. An iron boat was carried from Zanzibar to Lake Victoria and reassembled. The subsequent tour of the lake proved, once and for all, that Speke had been correct about it being the Nile's source. Lake Tanganyika, Stanley discovered during the same expedition, gave rise to another of Africa's great rivers, the Congo. 'Defeated at last,' says Maitland, 'the tenacious Burton reserved until 1881 his acknowledgement of Speke's ultimate victory and in 1890 wrote a letter from his death-bed in which he told Grant that every harsh word he had ever uttered against Speke was withdrawn.'

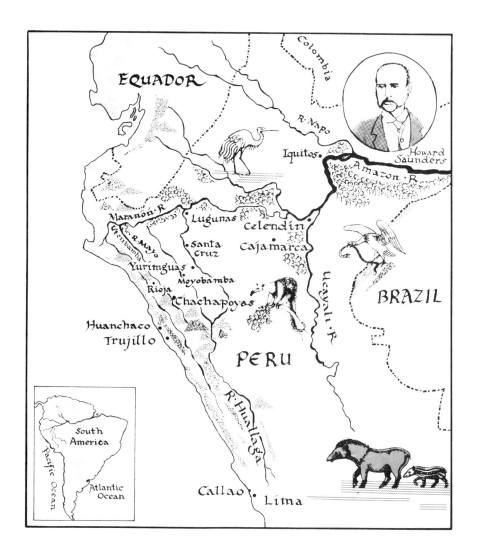

Howard Saunders's journey across the Andes, 1861–2

CHAPTER SEVEN
Rambles in the Andes

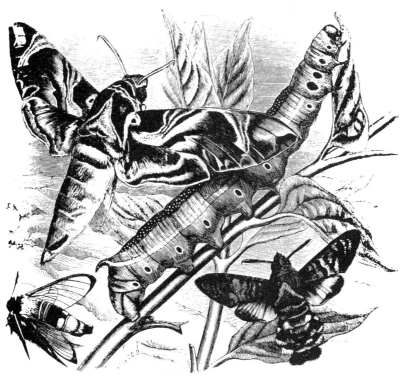

7.1 At night Saunders collected moths around the lamps in his camp including these hawk-moths

Howard Saunders is a good example of a field naturalist turned closet naturalist; he started out as a collector and traveller, and ended his life as a respected figure in English ornithology. As a young man Saunders travelled extensively in South America, and later in Spain and other parts of Europe. Wherever he went, he shot specimens for his large collection of bird skins, but he was also interested in other aspects of natural history and wrote detailed and lively accounts of his journeys.

Saunders's most important contribution was made through his work for various zoological and ornithological societies and journals. One of his friends summarises it thus:

It is impossible to estimate too highly the value of his life's work in the cause of Palaearctic Ornithology, but he did not devote himself solely to the study of birds, for he took the deepest interest in geographical research, more especially in that relating to the Arctic and Antarctic regions.

Soon after his return from collecting in Spain in 1870, Saunders was elected a member of the British Ornithologists' Union (BOU), and for the last six years of his life he acted as its Secretary. In 1892, he became the first Treasurer and Secretary of the newly formed splinter group of the BOU, the British Ornithologists' Club. From 1877 to 1881 he was the recorder for the *Aves* section of the

Zoological Record; and for five years from 1880 he was the Honorary Secretary to Section D (Zoology) of the British Association, a post held by Bates during John Hanning Speke's lifetime (see chapter 6). He also made good use of his editorial skills. With P. L. Sclater, another friend of Speke's, Saunders edited two volumes of the BOU journal *Ibis*. In addition to this, he was a Fellow and council member of the Zoological Society, the Linnean Society and the Royal Geographic Society, and published papers in their journals.

The experience Saunders gained from seven years in South America and as many years in visits to Spain enabled him to specialise his knowledge, and to devote most of his time to ornithology. It seems extraordinary that someone so immersed in the traditions and institutions of the English scientific world should have gone through the incredible hardships and dangers of Saunders's early journeys.

Howard Saunders was born in 1835, and his South American trip was made at about the same time as Speke was exploring in East Africa. Saunders was brought up in 'an old and honourable merchant family of the City of London', and attended schools in Leatherhead and Rottingdean. Naturally, as a young man he joined one of the merchant banks with which the family was involved, Anthony Gibbs and Sons in the City. In 1855, at the age of twenty, he set out on the clipper *Atrevida* for South America, apparently on the recommendations of certain of the bank's clients. He spent some time in Brazil and Chile before arriving in Peru where he passed the next four years 'in antiquarian researches, and in acquiring a perfect knowledge of the Spanish language'. Peru at that time 'offered to an explorer, and particularly to an ornithologist, magnificent opportunities of which Saunders was not slow to avail himself.'

Saunders decided to return home by crossing the Andes and following the River Amazon down to the Atlantic. This was something of a long-standing ambition since, as he puts it, 'the day when, five years previously, I first set foot in Brazil and saw the glories of its primaeval forest, the idea of reaching that country from the western side had never been lost sight of.' He discovered a little bit about the route from the writings of William Smyth and Frederick Lowe, two lieutenants of the British Royal Navy, who had travelled through the Andes in 1836 (see chapter 4); and from the reports of a handful of other travellers who had made the journey more recently. Saunders arrived on the headwaters of the Amazon only a year after a fellow naturalist, Henry Bates (see chapter 5), left for England and must have been fascinated by Bates's *The Naturalist on the River Amazonas* which appeared the year after he got home.

The coast of Peru is notoriously arid, but there are two rainy seasons in the high sierras. Hoping to avoid the worst of these, Saunders left Callao, on the coast near Lima, by steamer in April 1861 and travelled 300 miles north to Huanchaco. Moving inland to Trujillo he found 'a clean bright city with the streets built at right angles, and inhabited by more Spanish-blooded people than any other place in Peru; it is, in fact, the aristocratic stronghold.' The countryside was fertile with sugar growing in the valley bottom and a local industry based on cochineal insects. Cacti were grown as their food plant and the insects were used to make red food colouring.

By the time mules had been obtained and baggage arranged, it was early May, and on the 10th Saunders set out for Cajamarca, 60 miles northeast. He spent a night with a friendly Irishman who crammed gifts into his saddle bags until the mule was so loaded down that Saunders, for the first and last time as he firmly points out, had to use a chair to mount. But this was not all:

> As he kept pressing me to have more, I remarked that there was nothing wanting but a warming pan, on which he exclaimed 'There's one in the house this minit; will ye have it?' This was too much. I kicked my legs vigorously in a vain attempt to reach the flanks of my beast with the spurs, and the last I saw of my generous host was standing

in the doorway, waving his hat with one hand and brandishing the warming pan in the other. How the thing got there, Heaven alone knows!

The road started to climb through a series of passes into the mountains. The only living thing Saunders saw was a large fox scavenging at the corpse of a horse by the wayside which had died the previous day and was already picked clean by the ubiquitous turkey vultures. He stayed that night in a village of houses with overhanging eaves and gardens full of orange and custard-apple trees. At the top of one of the higher passes, above the vegetation line, Saunders shot a brace of what he describes as a partridge, actually a species of wood-quail. Descending into the next valley he learnt that when going downhill 'there was nothing for it but to sit well back and let the mule go; for they dislike being led and are as likely as not to fall atop of you.'

Darkness fell as Saunders's party was still making its way to the Indian village where they were to stay. Fireflies dazzled them as they pushed their way through the thick vegetation in the valley bottom. Suddenly they came into a blaze of red light. A simple sugar mill was being turned by oxen and the glow from the fire burning up the sugar cane stalks lit the dark forest pressing in on all sides and silhouetted the Indians feeding the mill.

When climbing a narrow trail, the mules were always securely tied one behind the other with ropes attached not onto the saddles but knotted on their tails. The need for this became clear to Saunders when he saw one animal lash out with its heels, lose its balance and fall over the edge of a precipice. And there it hung struggling in vain to get a footing but saved by the other mules in the train which stood firm until it could be hauled back onto the path. Soon they were able to look down on the basin of Cajamarca, over 10,000 ft above sea level and one of the most fertile valleys in the Sierra. The red-roofed town surrounded by pastures full of cattle reminded Saunders of English countryside, but for the towering mountains and plumes of steam rising from the hot springs. It was here that the Inca Atahuallpa had been living when Pizarro and his Spaniards crossed the Andes over 300 years before.

On the plain surrounding the Inca springs outside the city were flocks of 'lliclic plover' or killdeer. 'This species is known as the Christian or orthodox bird, because when asked 'Where is God?' it jerks it head upwards three times, intimating that He is in Heaven and there are Three in One', Saunders explains. Because of this, the local people would neither kill nor eat the plovers although the birds were very tame.

The marshes round the warm pools were full of shrill-voiced Andean gulls and great egrets stalking through the shallow water. A larger spring nearby was used to scald pigs. Legend had it that much of Atahuallpa's gold had been thrown into the pool to keep it from the Spaniards. As a result Saunders noticed: 'The moment a visitor approaches and peers into its depths the pig-scalders knock off work and follow every movement, ready to cry halves in anything that may be discovered.'

Leaving Cajamarca was no easy task. 'If any man thinks that money will do everything, let him come and undeceive himself in South America', wrote Saunders in exasperation. He could find no muleteer who was prepared to go down into the Marañon river gorge and cross into the eastern sierras with him. Eventually he joined a mule train belonging to Don Santiago, an old friend and landowner from Chachapoyas which lay to the north and east of Cajamarca. He was delighted to see the gun on Saunders's saddle since his own mule was carrying a conspicuously large bowl of silver and other silver utensils strapped to its pack. While fording a stream, Saunders shot a couple of glossy ibises which they had for supper although the birds were not as good to eat as the killdeer which he had shot, surreptitiously, near Cajamarca.

They crossed into the next valley which was well cultivated like Cajamarca but 1000 feet lower and therefore more pleasant in climate. The native Indians and Mestizos of the town of Celendin where they stayed were so friendly that, as Saunders puts it, 'they nearly danced us off our legs in their anxiety to practise the dances of the civilised world, and by hiding our mules they forced us to remain there a day longer than we had originally intended.' In Celendin, Saunders met an Irishman from Trinity College, Dublin, who had come out to Peru as a private tutor and was making a living by buying sugar and other produce in the warm, lower valleys and trading them in the higher villages and towns.

Climbing up from the valley once more, Saunders and Don Santiago reached the next ridge as the sun rose in the east ahead of them.

We stood on the edge of a vast cauldron filled with seething mist, which gradually melted before the rays of the sun, disclosing the blue range of the true Andes, with their fantastic peaks in full relief against the sky, whilst at out feet, 7000 feet below, the Marañon, the infant Amazonas, looking from this distance like a silver thread, cleft a passage through the Cordillera and the eastern range.

They began the long descent on a narrow path which zigzagged its way down the mountainside. At times teams of mules carrying blocks of salt or sacks of produce pushed past impatiently on their way up the trail. It was a battle of nerves, and Saunders advises all would-be Andean travellers to carry a revolver in case it becomes necessary 'to drop' an over-loaded mule threatening to push you over the edge.

They re-entered the vegetation zone and saw bright green Amazon parrots in the brushwood. Saunders collected some specimens of an interesting snail with a transparent shell curled into long whorls. Later he discovered that the Indian to whom he had given the snails for safekeeping had promptly eaten them, but he assumed the species had been one already described by an earlier collector in Peru. When they reached the Marañon, the greatest river of the Peruvian Andes which flows from north to south, they found a rushing, foaming torrent emerging from a steep gorge. The mules were unsaddled and swam across guided by one of the Indians. Saunders and his companions crossed the river on a balsa or raft, ending up far down stream. The mountains towered over the river, some with their peaks in cloud, so it felt to Saunders as if he was looking up out of a deep well at the sky.

They sat talking with the ferryman that night while a puma 'began its hideous caterwauling away on the mountainside'. Although it was the eve of the feast of Corpus Christi and the village was resounding with firecrackers and music, Saunders decided to sleep out with his gun in the hope seeing the puma.

It was three in the morning, as I found out afterwards, when I was awakened by a terrific uproar of squeals and whinnying, and I had hardly sat up before I was knocked over by one of our beasts, all of which had been stampeded. As the cloud of dust they had raised subsided I saw by moonlight, and guided by the noise, the puma bounding up the rugged mountainside with a pig; and the pugs afterwards showed that he had taken it from only a few yards from where I was lying, and from right under the noses of the mules and horses.

The path which climbed out of the Marañon gorge was in places less than two feet wide. Here it was neceesary to dismount and, much as they detested it, to lead the mules for 'a touch of the foot or stirrup against the rock would have sent man and beast over a sheer precipice some three thousand feet or so.' Saunders carefully measured the width of the path. With his feet placed toe to heel, the sole of his outer foot projected over the edge. They descended safely to the valley of Illabamba where the inhabitants, he learnt, paid their taxes in gold mined from the surrounding quartz. The fields of maize and peas around the village were full of deer, and flocks of pigeons and green mitred parakeets with brilliant scarlet throats and foreheads.

Saunders helped the villagers herd their cattle into pens to be branded, nimbly avoiding the bulls which charged the first thing they saw on entering the village. The beef was the only meat available and was stored by drying. When boiled it resembled leather and, if baked, it was so hard that it cut the gums. However, there were plenty of vegetable foods in the form of potatoes, sweet potatoes, yuccas, various peppers and avocado pears.

Saunders's party left Illabamba in mid June and climbed into the eastern sierras. They reached the first pass, 15,000 feet above sea level, in a blinding hailstorm and gale. The way down proved difficult. 'I had anticipated that the roads would get worse, yet this bit fairly surpassed my expectations', wrote Saunders. The road led over steeply sloping wet rock and through quagmires on a kind of 'corduroy' surface of tree trunks laid by the Indians. The mules occasionally fell into mud up to their girths. They stayed in a poor village under leaden skies and it rained hard that night. In the morning the mules who had taken advantage of the mud for a good roll, 'were more than unusually unpleasant to saddle'. Saunders and Don Santiago followed a tributary of the Marañon for most of the next day, disturbing herons and bitterns from the waterside and a noisy 'blue and yellow jay', probably the species known as the green jay. They crossed tributaries of the main river on wooden bridges, roofed in and gated at either end. The river passed through a narrow gorge with walls over 3000 ft high which were swarming with chestnut-collared swifts.

Two days later the men were approaching Don Santiago's home town of Chachapoyas on an old Inca road. This had been skilfully shored up where it skirted the precipices, Saunders noted, and was still in good condition despite three centuries of neglect. They stopped to rest and quench their thirst in the village of Magdelena where the only drink available was guarapo, fermented sugar cane juice. As a result, they continued their journey with what Saunders describes as 'a very cheerful view of things in general'. It was almost dark when they finally reached Chachapoyas. The square was filled with a candle-lit crowd to welcome them while the church bells clanged and rockets were let off into the night in true festival tradition.

Saunders had the usual trouble in finding guides and porters for his onward journey. As things turned out, this was most fortunate for he was still delayed in Chachapoyas when six men of the Aquaruna tribe arrived in the town. They carried blow-tubes 6–10 ft in length, and poison-tipped arrows. The leader of the group, to Saunders's delight and intense interest, was wearing a magnificent chaplet of hummingbirds. In the centre of this was one of South America's rarest species, a marvellous spatuletail. This extraordinary little bird, with a body only 4–5 in. long, has a forked tail with two central tail feathers narrowed and elongated to about 12 in. These feathers cross over in a semicircle and are tipped with large bright purple rackets. The hummingbird's body is no less spectacular as its whole plumage is brilliantly and glitteringly coloured. The crown is purple, the throat blue edged with green, and the back bronze green.

The marvellous spatuletail is known only from the Chachapoyas region of Peru, and the entire species is thought to be confined to the east side of the valley of the river Utcubamba, a tributary of the Marañon. It was first discovered by an English botanist named Mathews in 1835, 26 years before Saunders's visit to Chachapoyas, but it had never been collected since. Afterwards, Saunders could not forgive himself for failing to obtain the hummingbird from the chieftain. It was not until 1879 that a second specimen was secured, this time by a Mr Stolzmann who was collecting for the Warsaw museum.

To Saunders's frustration, he was forced to choose a route for his onward journey which led due east to Moyobamba, 'a line only just practicable for quadrupeds, and a large portion of which had virtually to be traversed on foot.'

Not long after leaving Chachapoyas, Saunders came upon a king vulture on a dead mule, surrounded by the smaller black vultures waiting at a respectful distance. He was particularly interested to see so many black vultures high in the Andes whilst the turkey vultures, so common on the coast, he noted, were absent.

Saunders had been joined for this stage of the journey by a man he refers to as 'W', whom he describes as

> a huge German, who was travelling about south America for his pleasure, and whose company I had admitted with some reluctance, as he was quarrelsome, and presumed too much upon his size; however, he managed to die a natural death at Pará, poor fellow! after a more prolonged existence than could reasonably have been expected under the above conditions.

Together they climbed a formidable pass, spoken of with awe in Chachapoyas and named Piscu-huanuni or 'the place where the birds die'. Saunders, however, seems to have taken it lightly and only reports that he 'saw a pipit on the very summit'. They descended into the rain forest on the eastern flank of the range and were soon huddled under waterproofs and Panama straw hats but passing through magnificent forest scenery. At night, any naked flames were immediately swamped in insects. Saunders amused himself by counting the different kinds of 'gigantic moths and innumerable hoppers and creepers' which came to a candle, but became confused after reaching 70 species.

Following the river Mayo they arrived at Rioja, the first village they had seen for many days. For want of a better guide in the deserted village, they followed a herd of pigs to a house. This, as it turned out, belonged to Don Crispino, the very man to whom Saunders had a letter of introduction! He made them welcome with cool grass hammocks, ale and Portuguese wine, and American crackers. Although reluctant to leave this luxury, the party headed on through forests full of gorgeous butterflies, and across a swamp until at last they approached Moyobamba. Despite the fertile fields surrounding it and a thriving local industry of straw hat plaiting, this was an untidy and inhospitable town. In addition to the inevitable swarms of biting insects, the town was plagued with snakes. And as Saunders points out, however much you like snakes, no one would enjoy,

> when standing one night by an old wall, to feel a long cold body — very long it seemed — gliding slowly over my neck, and slowly — oh, so slowly — finding its way to the ground. Doubtless from its size it was not a venomous snake; probably a rat-eating snake, about six feet long, but he seemed more.

The first night after leaving Moyobamba the party camped in a spot frequented by a jaguar. Nearby was one of its favourite watering places so Saunders took his gun, and a little Indian dog as bait for the jaguar, to lie in wait:

> The situation was not unfavourable, as a good view was obtained of the only place where the jaguar could come down to the water; but when the last light had left the sky a certain feeling of eeriness came over me, which was not conducive to a prolonged stay. Of course, the correct thing is to remain until, amidst the pitchy darkness, the gleaming orbs of the jaguar appear; you aim carefully between the glowing eyes, and, if the aim is straight — and it always is in books — you get your jaguar; otherwise he gets you, and the book is not written for want of an author. But how if that jaguar should all the time be crouching upon a branch above and behind you, making up his mind for a spring! The reflection was not altogether satisfactory; so, winding up the little dog into one prolonged and final howl, just to let the jaguar know that we were not in the least afraid, but really could not wait for him any longer, I picked up my living bait and made tracks for the camp.

However, that was not the end of the story because that night Saunders was kept awake by the roaring and chattering of a group of howler monkeys near the camp, stirred up to a frenzy by the vibrating roar of a jaguar. The next morning they found jaguar footmarks all around the camp, and in the sand by the river where Saunders had waited.

Despite the difficult terrain and regular drenchings in the tropical rainstorms or when crossing rivers, Saunders tells us he was in better health than at any other time in his life. In this he was more fortunate than many other explorers, and it must have accounted, at least in part, for his amusement and enthusiasm at even the trip's most unexpected events.

Finally he reached the river Paranapura, one of the many tributaries of the Amazon, and embarked in a canoe. 'The river scenery was very fine, and as evening approached the air seemed alive with birds, principally parrots, seeking their resting places for the night; and by the margin of some wooded islands wallowed the capybara and coypu.'

River navigation was made even more interesting by the occasional enormous fallen tree which ensnared the canoes. The branches had to be cut away with machetes in a desperate attempt to escape before the current capsized the frail canoe. At night they stopped in riverside villages. One such household, on an island in midstream, was swarming with half-domesticated guinea pigs which were considered a choice dish. Saunders, like Charles Waterton (see chapter 3), found them very palatable, and 'much more delicate than rabbit'.

The river Paranapura flowed into the Huallaga and here they turned upstream to the town of Yurimaguas. This was the highest point on the Amazon river system that the regular steamer service, established during Bates's stay on the river, could reach. In Yurimaguas, Saunders and his German companion stayed with the governor, called Subauste, who according to Saunders 'was virtually banished to this remote district, and heartily glad to see any civilised beings.' He fed them on fish and river turtles which, although Saunders became extremely sick of them later on, were a welcome change of diet from hard, dried meat. The turtles were served up in a variety of ways to provide an entire meal:

> first in soup, with the pickings of the delicious little black paws, like glorified turbot-fringe; then scalloped turtle, surrounded by a pattern of its own eggs, and with palmitos, the heads of a species of

7.2 The meat of the capybara is said to taste like ham or bacon, and Saunders and his Indian guides frequently hunted it for food

7.3 Manatees lived along the upper Amazon River in Saunders's time

cabbage palm, to vary the monotony of the daily plantains.

No wonder he soon became tired of the oily taste! They also ate guinea pig, paca (a large forest rodent), peccary and turkey-like curassows.

Snakes were scarce in the town and as a result there were plenty of rats. They tunnelled into the earthen floor and walls of the house in which Saunders was staying, and fell with alarming thuds from the thatched rafters during the night. Anything small laid out to dry was in danger of being stolen, and Saunders had to retrieve several precious articles from rat holes using a hooked stick.

News arrived that the river was falling, and the steamer could come upstream no further than Laguna (Lagunas), a town about 60 miles away. Saunders and his German friend decided to accompany Subauste to Laguna by canoe, planning to fish and shoot along the way and camp on sandbanks. They left Yurimaguas on 5th August after a solemn mass had been celebrated in honour of their departure. Sabauste's housekeeper, jealous of their departure, had contributed to their farewell by attempting to poison them all with coffee the previous evening.

Shortly after leaving the town, the canoes entered a series of lakes running parallel to the main stream of the river. These were full of fish which the Indians proceeded to catch by a novel method. A canoe full of barbasco, a ginger-like root, was collected and crushed to pulp. When thrown on the water, the fish rose to the surface stupefied and intoxicated. Some of the fish weighed over 200 lb. Their flesh was firm and delicious, and when cut into slabs, salted and dried, it formed the staple diet of the Indians on the river.

Manatees also lived along the water's edge. These melancholy-looking mammals are believed to have been the inspiration for tales of mermaids although their closest relative is actually the elephant. The manatees were difficult to see as they fed by grazing on grass and other vegetation which drooped into the water. Saunders, quite rightly, scorned the suggestion he heard later from 'some French cabinet naturalists' at the Zoological Society that the animals came out of the water to feed. 'At night', he reports, 'the presence of the manatee is frequently indicated by a soft, deep sigh, which has a very weird effect when suddenly emitted in close proximity by an invisible utterer.' Even in Saunders's day the manatees had their problems. Manatee fat when boiled down, clarified and salted was sold extensively and profitably up and down the Amazon.

Another local mammal was the dolphin, remarked upon by visitors to the rivers of South America from Charles Darwin onwards. The species found on the upper Amazon, below the rapids on the Huallaga, was pale coloured, Saunders noticed, whilst lower downstream he saw two different species. One of these, he says, 'was first sent to Europe by the distinguished explorer Mr. H.W. Bates' (see chapter 5).

At night, the party camped on sandbanks exposed by the falling river which they often shared with flocks of screaming terns. These were yellow-billed terns which nest opportunistically on banks in river beds and lakes, and then migrate north to the Caribbean. Further down the river, Saunders also collected the large-billed tern. Black skimmers trailed their bills over the water surface in search of fish, and noisy macaws and toucans, known locally as 'preachers', flocked to roost in the safety of the island trees. Near one of their camping sites, Saunders found a huge harpy eagle so gorged with food that he approached it to within 50 ft, but fortunately had not the heart to shoot it.

The riverside forest was also alive with game, as Saunders's description illustrates:

> As we pushed our way through the snags and brushwood which blocked up the narrow channel, an occasional crash and a momentary view of a dark brown body betokened the presence of the capybara, and more than once the whispered 'danta' of the Indians showed that their keener perceptions had distinguished the tapir. Splash, splash went the alligators every instant from the mud banks on either hand, and deafening became the shrieks and maniacal laughter of the kingfishers (*Ceryla amazona*) from the boughs above, as if they were determined to put up every head of game for miles round. A heavy rush through the reeds, and a tapir broke cover, and, springing from the bank, was instantly lost to view in the muddy waters, unharmed by a snap-shot from my gun. An instant of silence followed, and the air was filled with cries from myriads of birds, aquatic and others; whilst a dense black cloud of wings in the direction of the lake made me aware that I had missed far more than the tapir by that discharge.

On one of the sandbanks in the river, Saunders found a flock of flamingos and white wood storks with bare grey heads and necks. That day's shooting was rewarded with a bag of ducks and geese, and a specimen of the rare hoatzin. This bird is best known for its nestlings' possession of primitive claws at the joints of their wings, used to pull themselves from the water onto overhanging branches. 'Its flesh', wrote Saunders in disgust, 'is not only unfit for food, but had also the most repulsive odour.' So bad was the smell, that he threw the hoatzin's body overboard rather than tolerate it in the canoe.

They tried their hand at hunting in the forest away from the river, but shot nothing. Saunders was reminded of the towering trees and scanty undergrowth later when he read Bates's description of the forest being like a vast cathedral of which the observer is the sole occupant, whilst the performing choir is all outside on the roof.

Starting by moonlight for a dawn hunting expedition, they had an extraordinary encounter with a jaguar. The front canoe disturbed a jaguar swimming back from an island where he had been feeding on capybaras, and managed to head him off from the bank. Then in Saunders's words:

> [the] bow paddle gave him a passing blow on the head, and the next instant that jaguar was in the canoe amidships, whilst six dusky bodies were seen taking headers into the water. Probably he took off from a shallow, or a submerged tree-trunk, for there was no effort of clambering. The jaguar walked aft, and, seating himself on his haunches, yawned in a lazy way, whilst the abandoned canoe went round with the current; the

7.4 Hoatzins have a number of unique features, and it is thought they may be an evolutionary link between modern and ancestral birds

raised stern came within jumping distance of the shore, and, pulling himself together, with one graceful bound the jaguar disappeared behind the dense wall of foliage.

The party pushed on downstream to Santa Cruz, a village they found perpetually humming with mosquitoes, and finally arrived at Laguna. Here they waited awhile for the steamer and had almost given up hope, planning to abandon their baggage and travel on by light canoes on the falling river, when she arrived. 'How beautiful she looked at daybreak, and how delightful it was to swing one's hammock in the cool breeze under the awning!' wrote the travel-weary Saunders.

The steamer began her return journey downstream on 21st August, frequently running aground. Saunders landed briefly at Iquitos, at the head of the main Amazon. He crossed the border into Brazil just above Tabatinga, 750 miles from Yurimaguas. Even before this point, the river had become so wide that both banks were rarely visible at one time, and, Saunders admits, 'the scenery of the Amazonas, as viewed from the deck of a steamer, soon becomes monotonous.' So he travelled down the river to Pará on the Atlantic.

Saunders arrived back in England in 1862, but started travelling again the following year. For the next seven years, using his fluency in Spanish, he made himself an authority on the birds of the Iberian peninsula. Another reason for his choice, surprising as it may seem, was his deteriorating health. In 1867 he was advised to spend the winter in a warm climate, although Saunders's idea of convalescence was probably not quite what the doctor had in mind.

He arrived in Murcia on 4th November 1867, delighted to be back in Spain. The first thing he did, which took precedence over a cure for his crippling rheumatism, was to inspect the bird skins in the local museum. After a soak in the baths of Archena in the hills above the town, Saunders felt sufficiently restored to start observing and collecting birds. In the market in Murcia, he saw starlings for sale and was told that after being bled to remove their bitterness, they were delicious. He travelled on to Malaga, within 60 miles of Gibraltar, where he intended to spend the winter. Fortunately, as he reports, 'under its genial climate I gradually threw off my rheumatism, and became as well as ever I had been in my life.'

The estuary of the Guadalorce river close by provided splendid birdwatching and the opportunity to collect more specimens. In addition, the Malaga market sold red-legged partridges, golden plover, stone curlew, and a few black-tailed godwit, grey plover and dotterel. Among the songbirds on sale as food were skylarks, cirl and ortolan buntings, crested and calandra larks and a pile of sparrows from which Saunders was able to extract a Spanish sparrow and a rock sparrow.

The winter of 1867 – 8 was unusually severe. Saunders, rheumatism forgotten, accompanied a wolf-hunting party up into the mountains. The shots and cries of the beaters put up scores of vultures and Bonelli's eagles. In early February, Saunders took the steamer to Cadiz, and watched hundreds of gannets plummeting after fish in the Strait of Gibraltar. He travelled north by train to Seville, passing through meadows full of storks and cranes. By mid March he had reached Granada, where he was disappointed to find heavy falls of snow still blocking roads into the Sierra Nevada in which he had hoped to see the rare lammergeiers, a species of vulture almost extinct in Europe.

By this time, Saunders felt in need of another session at the baths of Archena. With typical foolhardiness, he decided to travel cross country to Murcia. Starting off from Granada in a stagecoach which left at one in the morning, he found himself high in the gorges of the Sierra Nevada by day break, and was rewarded by magnificent views of a family of lammergeiers. The last part of his journey was completed in what Saunders describes as 'the very fastest diligence in which I ever travelled'.

As soon as his health had been restored at Archena, Saunders set out back to Malaga, hoping to see the spectacular spring bird migration across the Strait of Gibraltar. He also wanted to collect the much coveted eggs of the rare Egyptian vulture. Descending the breeding cliff on a rope, Saunders reached the vulture's nest and found two fresh eggs. These, some of his friends assured him later, if eaten would ensure his survival 'to the age of Methusaleh'.

He hurried on to Seville in time to search for nesting bustards on the plains nearby. The birds were very wary and even night-time stalking did not bring any of them within range of Saunders's gun. The wild cattle, on the other hand, were aggressively friendly; as Saunders puts it: 'At the best it is nervous work to find yourself the observed of some 200 cows, each watching jealously over her calf.'

In late April, Saunders and a friend called Manuel set out for the river marshes of the Coto del Rey and Coto de Donaña. Here they continued their raid on birds and nests alike, adding bitterns, purple herons, egrets, pratincoles, stilts, hoopoes, kites, harriers, spoonbills and many other species to Saunders's growing collection.

After returning briefly to Seville, Saunders dashed off once more on hearing that a Bonelli's eagle nest had been found in the mountains. This he reached on a rope, using a stick to push himself out from the sheer rock face and then swinging in under the overhang to the nest ledge. He returned triumphantly to the cliff top with two live eaglets tied up with his braces, and a mysterious egg which appeared to have been laid by a bird brought to the nest by the eagles as food. The young eagles, Saunders decided, should be taken back to London and donated to the Zoological Society's gardens, now London Zoo.

The duration of his stays in Seville was becoming shorter and shorter. After only an hour in the city, Saunders was called off to a trial of young bulls for the bullfighting ring which was held on the plain beside the Guadalquivir River. Nearby he succeeded in finding four bustard nests complete with eggs which were promptly added to his collection. Saunders planned to return home by steamer with his eaglets, and his remaining hours in Seville were now numbered. Manuel paid a last, flying visit to the swamps near the city and returned with a huge basket full of little egrets, squacco herons and glossy ibises for Saunders.

'We were hard at work skinning till past midnight,' he recalls, 'and at 6 a.m. I was on my way to Cadiz to join the steamer for London.' So ended Saunders's winter of convalescence in Spain.

Upon his return to England, Saunders married Emily Bigg and settled in a house near Hyde Park in central London. His marriage is said to have proved 'an exceedingly happy one, for his wife took the keenest interest in his work, and the help which she afforded him in his scientific career cannot be too highly spoken of.' And it was upon a scientific career that Saunders now embarked. In 1869, he published a lengthy account of his Spanish collecting trip entitled 'Ornithological Rambles in Spain' which appeared in the journal *Ibis*. This was followed by notes on the ornithology of Italy and Spain, and after his last visit to Spain in 1870, by a collection of papers on European birds. In 1881, he finally published the account of his journey across the Andes as a series of articles in *The Field, The Country Gentleman's Newspaper*. By the end of his life, Saunders had produced over 50 papers and written or edited four books.

Howard Saunders also provided valuable advice to fellow naturalists and ornithologists

7.5 A woodcut of a hoopoe from An illustrated manual of British birds *by Saunders (Courtesy of the Alexander Library, EGI, Oxford)*

for whom, it is reported, he kept an open house. He used his experience and talents for meticulous record keeping on numerous committees and as editor of *Ibis*. He maintained close contact with other travelling naturalists, and published papers commenting upon the birds collected by various expeditions, including those of HMS *Challenger* and the Transit of Venus expedition to the south Atlantic in 1874–5. He became a recognised authority, in particular, on gulls and terns. In 1896, Saunders produced a monograph on the gulls and terns, the so-called *Gaviae*, in the British Museum collection.

Despite all this book work, Saunders continued to travel. In 1883–4 he visited the Pyrenees, in 1891 he was in Switzerland and two years later in France. In 1897, when Saunders was nearly 70, he and his wife and two daughters spent a holiday in Norway with Abel Chapman. In Chapman's words: 'Through long practice, both at home and amid the denser jungles of southern lands, he had acquired remarkable quickness in identifying small species in the open, even though but half-seen among foliage or reed-growth.'

By 1906, Saunders was fatally ill but intent upon producing the third edition of his *An Illustrated Manual of British Birds* which had been published first in 1889. Howard Saunders's death in 1907 occurred within months of that of Alfred Newton, and both men were mourned by the scientific community. To quote one of his obituaries:

> These two men were universally acknowledged to be our most learned authorities on British Birds. All difficult questions relating to British ornithology were invariably referred to one or other of them, and no one every appealed for help without obtaining the fullest information and soundest advice.

From the precarious mule tracks of the Andes to the peak of the ornithological world of his time, Saunders had made his contribution as one of the most versatile of the travelling naturalists.

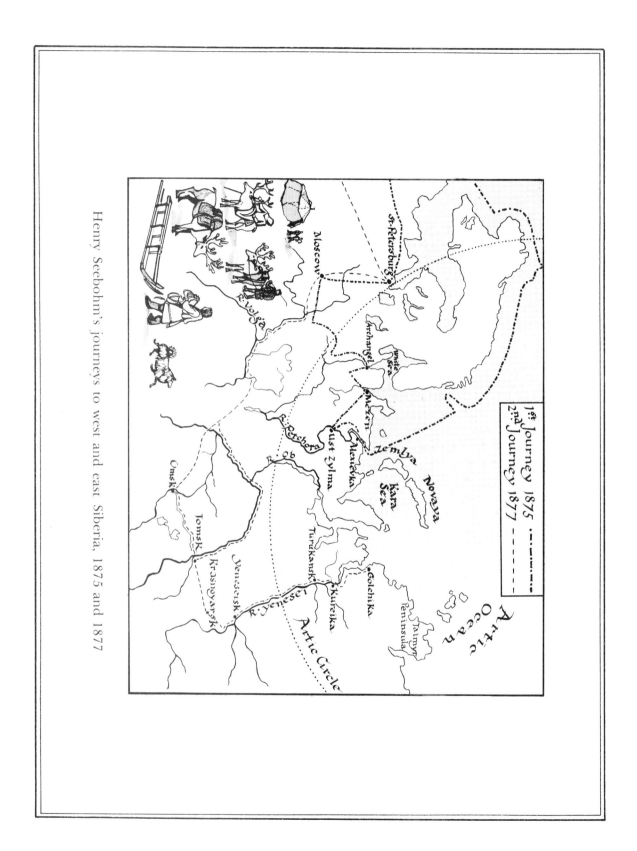

Henry Seebohm's journeys to west and east Siberia, 1875 and 1877

CHAPTER EIGHT
To the Yenesei and Petchora

8.1 Henry Seebohm, ornithologist and collector, in an engraving made from a photograph (Courtesy of the Alexander Library, EGI, Oxford)

Nearly half way across Russia's vast breadth lies a range of mountains called the Urals which run from north to south and divide northern Russia both geographically and faunistically into east and west. Either side of the Urals, a great river flows northwards into the Arctic Ocean. On the west, the Petchora River rises from the foothills of the Urals. To the east is the much longer River Ob which drains the vast shallow Siberian basin from its headwaters in the southern plateau north of the Aral Sea and Lake Baikal. Even further to the east, a third major river, the Yenesei, empties into the Arctic Ocean. Beyond its estuary the frozen wastelands of the Taimyr Peninsula stretch away to Russia's most northerly tip.

During the first decades of the seventeenth century, the Petchora River valley was visited by a number of English merchants who established a fur trade in the region based upon the rich beaver pelts they found there. The next visitors, from St Petersburg (Leningrad) and Sweden during the first half of the nineteenth century, collected some information on the geology, botany and ethnology of the region. When the ornithologist Henry Seebohm travelled to the Petchora in 1875 (and two years later to the Yenesei), he and his companion were the first Englishmen to visit the area for 250 years.

Henry Seebohm was born in Bradford, Yorkshire, in 1832. His family were Quakers and, although earlier resident in Germany for several generations, they had come originally from Sweden. Seebohm was sent to the

Friends' School at York where he showed a particular interest in natural history and made a sizeable collection of ferns, birds and eggs. His next years were occupied with building up a steel manufacturing business in Sheffield, a profession which he retained all his life.

Seebohm loved to travel, and each year he made short trips to different parts of Europe and once as far as Turkey. His main purpose on these journeys was collecting birds, a pastime for which he had a particular fascination. He added to his schoolboy collection of birds' skins and eggs throughout his life. Although he rarely missed an opportunity to obtain a new species, Seebohm specialised mainly in three different groups of birds: thrushes, wading birds, and warblers in the genus *Phylloscopus*, whose almost identical appearance presented special problems for anyone attempting to identify the many different species.

Seebohm made his first visit to Siberia in 1875. It was a region which had particular interest for him since little was known about its birdlife and, divided by the Urals, it held representatives of both European and Asian avifaunas. All through his account of this and his later journeys Seebohm refers to the beauty of the scenery around him and carefully describes the birds he saw but adds quite naturally exactly which ones he was able to shoot and how easy it was to bring them down. He was obsessed, above all else, with collecting, and took as many species and as many specimens of each species as fell within the range of his guns. By today's standards, this seems wasteful and improper, but at least Seebohm was able to put the information he gained to good use. Using his enormous collection of skins and eggs, he became the first to establish the exact distribution of many Siberian birds. On his death, Seebohm's specimens were donated to the British Museum and formed the basis of what is now one of the world's larger ornithological collections.

Seebohm set out for Siberia with his friend and fellow ornithologist John Harvie-Brown. They travelled light, each taking only a change of clothes, rubber boots, mosquito-proof veils, tent, hammock, a breech loader and a walking stick gun. By the time they reached St Petersburg it was exceedingly cold; the snow was two feet thick and sledges were the only means of transport in the city. Like naturalists all over the world, the two men first examined the city's food markets carefully for samples of the local birds. They were rewarded by the purchase of a dozen fine waxwings which on dissection turned out to be all males. Since he knew that the birds gather into flocks in winter, Seebohm concluded that in waxwings, as in some other species, the sexes are segregated in separate flocks.

The men moved on to Moscow, and then to Vologda 250 miles further north where they equipped themselves with fur dresses, provisions and a sledge. They left Vologda at eight o'clock on a Sunday morning and travelled day and night arriving the following Thursday in Archangel, having covered a distance of about 550 miles. The sledge was drawn by successive teams of three horses which Seebohm describes as 'tough, shaggy animals, apparently never groomed, but very hardy'. The whole journey used 108 horses; the drivers urged the team on with shouts and oaths, sometimes imitating the bark of a wolf to frighten them into galloping faster. The countryside through which they passed, still rigid with winter, had few birds or animals although Seebohm saw ravens, jackdaws and some finches.

Archangel had a population of about 16,000 which doubled in size during summer. The houses were built of wood with spacious gardens, apart from the centre of the town which had plastered brick buildings. Seebohm and Harvie-Brown spent nearly three weeks in Archangel making more preparations and inquiring about the Petchora valley that lay ahead of them.

Near the town they found some Samoyedes, a nomadic Mongolian race, in an encampment of tents called chooms, made from reindeer hide stretched over birch-wood frames. The Samoyedes were small and had long straight black hair. Their features were typically Mongolian with flat noses, high cheek-bones, thick lips and slanting eyes. Unlike the two Englishmen, the Samoyede men had no side whiskers and their moustaches and beards were thin. Seebohm took some of the Samoyedes into the local museum where they recognised many of the stuffed birds and gave imitations of their calls and behaviour so that Seebohm was able to learn some of the Samoyede names.

This was Harvie-Brown's second visit to Archangel. On his first visit with E. R. Alston in 1872, he had met a Polish exile named Monsieur Piottuch who had an interest in birds. The two men now found Piottuch again, and engaged him 'in the double capacity of interpreter and bird-skinner'. The weather was mostly cold, dry and sunny but by the beginning of April, the temperatures were rising and, fearing that a thaw would make the roads impassable to sledges, they hurried to complete their preparations.

Their next destination was Ust-Zylma on the Petchora, about 750 miles away. For two months in spring the road turned to an impassable swamp and the valley was cut off from the outside. Seebohm found that they had left Archangel only just in time. The road surface was already thawing and was split into deep ruts by the busy traffic to and from a fair in one of the towns ahead. Most of the way led through spruce forests where they saw snow buntings, capercaillie and occasionally a Siberian jay. They paused in Mezen and watched the village boys catching snow buntings with snares of horsehair. As Seebohm notes disapprovingly, he and Harvie-Brown were glad to leave Mezen 'as Piottuch's old friends were too many for him, and far too hospitable, and he was drinking more champagne than we thought prudent.'

The three men travelled on through forests, river valleys and open plains. Although occasionally delayed when horses were not available, they generally made good speed with the return of cold weather. When at last they reached the Petchora the great size of the river impressed them beyond all expectations. The town of Ust-Zylma they found to be 'one vast dung-hill'. The peasants threw out piles of rubbish and manure from the stables beneath their houses so that, when the spring thaw set in, raised sidewalks were needed in the town's street to avoid the pools of liquid manure. The farming was very poor, but in summer the people supplemented their diet with salmon netted from the Petchora.

In Ust-Zylma Seebohm met Engel, 'a wild, harum-scarum, devil-may-care fellow', who was the captain of a steamer belonging to the Petchora Timber-trading Company, and he provided plenty of information about the surrounding country. They bought snow-shoes and started looking for birds outside the town. Apart from the numerous snow buntings, few other birds had returned so the list remained disappointingly short, with only nine species. Now they were safely in the Petchora valley, Seebohm was anxious that the spring should arrive quickly, and the birds with it.

The temperatures remained low and fresh falls added inches to the snow drifts which did little to cheer Seebohm and his friends. They collected specimens of the few birds around the town, and turned their attention to the Samoyedes. Some of them were reported to be camping at a village nearby and, on Engel's advice, the men paid them a visit. The Samoyedes spent their lives moving with their reindeer between the rich grazing of tundra to the north in summer, and the spruce forests and plains further south in winter. When mosquitoes made the latter unbearable to man and beast in spring, the Samoyedes packed up their tents and moved off with their herds.

The chooms had already been taken down and stowed on reindeer sledges when Seebohm arrived in camp. Using dogs resembling Pomeranians, some of the men were separating out 50 reindeer belonging to a local Russian from a milling herd of 500 animals. The reindeer to be left behind were lassoed skilfully by the Samoyedes and hobbled by their owner. The scene was one of complete chaos. The reindeer thundered to and fro herded by overexcited dogs, accompanied by a sledge driven at breakneck speed, bearing the owner who was trying to identify his animals. Sometimes the herd broke through the ice and floundered in deep snow drifts. The dogs tied to the waiting sledges barked furiously, and the reindeer that had already been caught hobbled and tottered frantically. After the day's march, Seebohm and his companions were invited inside one of the newly erected chooms. Protected from draughts by the snow piled around the outside of the tent, and warmed by a fire in its centre, they drank tea and listened to descriptions of Samoyede customs.

By the beginning of May, the thaw was well under way and the men were able to make long excursions on the banks of the Petchora. Migrating birds began to appear; first some gulls, then a single redstart, and later a white-tailed eagle perching on an ice block in the middle of the river. Seebohm lay in wait by the river for the first flocks of geese on their way to breeding grounds in the far north, but succeeded in shooting only one bean goose.

The ice on the Petchora River broke up and the streets of Ust-Zylma became awash with melted manure. The flocks of ducks, geese and swans followed each other in procession up the river valley and the woods around the town filled up with redpolls, meadow pipits, field-fares, redwings, lesser spotted and three-toed woodpeckers and marsh tits. Seebohm finally succeeded in shooting one of the gulls from a flock on the river whose identity had been puzzling him. He concluded that it belonged to a new subspecies of the familiar European herring gull, and called it a Siberian herring gull. In fact, the herring gull is a species with complicated taxonomy and a circumpolar distribution stretching from northern Europe to the Canadian sub-Arctic. A series of subspecies overlap each other in a necklace pattern across the herring gull's range.

At the end of May, Seebohm added another interesting bird to his collection. He shot one of the warblers living in willow scrub near the river and found, despite his vast knowledge of *Phylloscopus* warblers, that it was a species with which he was unfamiliar. The bird was a Siberian chiffchaff (now recognised only as a subspecies), and turned out to have been described first by a Swedish collector some years previously. On the invitation of Captain Engel, the three men travelled 25 miles downstream on the flooded Petchora to Habariki. They collected several new species of songbirds during their three day stay there and saw a variety of waders. These included newly arrived golden plovers, snipe, green-shank, stints, and other wading birds.

A sailing boat with a small cabin was found for the journey to the river mouth, and they set off on 10th June. The strong current carried them northwards but they stopped regularly to collect birds and eggs, for the breeding season was now underway. Just north of the Arctic Circle, Seebohm collected a strange pipit on one of the midstream islands. To this were added four more skins and, on their return home, the specimens were examined by Henry Dresser (see C3 and 4). He decided that the species was a new one and named it after Seebohm. Later Seebohm discovered that the bird had been described in 1863 by Swinhoe who had collected the pipit in China whilst it was on migration. So little was known about the Siberian pipit, as it was called, that Seebohm regarded it as one of the most interesting finds of his trip. The species, he concluded, nested on the tundra from the Petchora valley eastwards to the Bering Straits. After breeding, the birds migrated east and then south via China to winter in the Malay archipelago.

Before long the boat had entered the huge delta of the Petchora River, 'a labyrinth of water and islands, one almost as dead flat as the other' as Seebohm describes it. They collected more wildfowl and waders, and found it was light enough for them to shoot through the night hours. After ten days on the river they reached Alexievka, on an island in the delta, which was the port of the Petchora Timber-trading Company. Among the wader species which Seebohm had particularly hoped to see and collect was the grey plover whose breeding grounds were known to be confined to the tundra. At Alexievka, for the first time, they were truly in the tundra zone:

> We looked out over a gently rolling prairie country, stretching away to a flat plain, beyond which was a range of low rounded hills . . . It was in fact a moor, with here and there a large flat bog, and everywhere abundance of lakes. For seven or eight months in the year it is covered with from two to three feet of snow. . . . The vegetation on the dry parts of the tundra was chiefly sedges, moss, and lichen, of which the familiar reindeer-moss was especially abundant. In some places there was an abundance of cranberries, with last year's fruit still eatable, preserved by the frost and snow of winter.

They crossed the tussocky ridges of moor and staggered through the black swamps in the dips between them, glad of their rubber boots. To their great excitement, a small flock of grey plovers were seen and, thanks to diligent searching by the Samoyede guide, they found a grey plover nest, the first ever discovered in Europe. It was quickly relieved of its clutch of four eggs, and the bird shot as it returned to the nest, 'all in the cause of science'. By the time they returned to Alexievka at midnight they had collected fourteen grey plover eggs. The eggs were carefully blown and the shells added to the expedition's collection; the contents, only slightly incubated, made 'an excellent omelette' for breakfast the next morning. The local workmen had been offered a reward for finding nests, and the bag for the day totalled 162 eggs of 12 species.

Next the expedition moved downstream to Stanavialachta. Here they collected more waders including grey plovers and their eggs. Seebohm tried his best to obtain one of the swans nesting nearby, hoping to confirm that they were the rare Bewick's swans. He had been sent a clutch of eggs by a peasant who claimed to have the skin of the bird which laid them. Since attempts at shooting or trapping the swans had served only to cause the birds to desert their nest, Seebohm sent a servant off to Alexievka to find the peasant and buy the mysterious skin. He returned triumphantly with the swan's skin and its beak which had been removed and given to the children of the household as a toy, but which now provided conclusive proof that the species was a Bewick's swan.

The mosquitoes had become very bad, as Seebohm describes:

> Our hats were covered with them; they swarmed upon our veils; they lined with a fringe the branches of the dwarf birches and willows; they covered the tundra with a mist. I was fortunate in the arrangement of my veil, and by dint of indiarubber boots and cavalry gauntlets I escaped many wounds; but my companion was not so lucky.

8.2 Seebohm was one of the first to find grey plovers breeding, and he brought the first clutches of their eggs back to Britain

To avoid the worst of the insects, the three men moved down to some islands behind a sandbar in the estuary. Here Seebohm shot another on his list of wanted species, a little stint. Although he had no time to search for nests he suspected the birds were breeding nearby. Later they were able to arrange a return visit to the place, and succeeded in finding little stints and their nests and young. On this trip, the weather became windy and the sea too rough for the steamer to collect them on time. For several days they were forced to eat duck's eggs, grey plover and fried dunlin to eke out their stores.

Their last few days on the tundra were spent at Alexievka. On 1st August, Seebohm and Harvie-Brown boarded the *Triad* with a cargo of larch and, narrowly avoiding running aground on the sandbanks of the Petchora estuary, they set sail for Elsinore in Denmark. The passage took 35 days and was rough and foggy throughout. Without any provision for passengers on the *Triad*, the men's food was soon used up and they had to subsist on 'hard captains' biscuits, Australian tinned meat, and coffee with no milk and short rations of sugar.'

The expedition had been a success. Seebohm had collected a huge number of different Arctic birds including some species previously almost unknown. In this he had been particularly fortunate in the choice of collecting grounds because, he concluded, the birds 'appear to be extremely local during the breeding season'. Eggs of three species, the grey plover, Bewick's swan and little stint, had been obtained for the first time ever, and breeding had been confirmed for the sanderling and knot, two species which had been baffling ornithologists. A total of 110 different species had been collected and several species added to the European bird fauna. Seebohm's notebooks contained records of first arrival dates and breeding seasons for many of the migrant birds, and other species he was able to confirm were resident all year round.

It was not long before Seebohm had the opportunity to return to Siberia and to enlarge his Arctic collection. This time he was to go even further east, to the valley of the Yenesei River. The area had been explored in the sixteenth century by a British expedition under Sir Hugh Willoughby. In Seebohm's words:

> Three ships were sent to the Arctic region on a wild-goose chase after the semi-fabulous land of Cathay — a country where it was popularly supposed that the richest furs might be bought for an old song, where the rarest spices might be had for the picking, and where the rivers rippled over sands of gold. Like so many other Arctic expeditions, this proved a failure. Poor Sir Hugh Willoughby, it is supposed, discovered one of the islands of Novaya Zemlya, but was afraid to winter there, and landed on the Kola peninsula, where he and all his crew were starved to death.

Another ship from this unfortunate expedition was blown into the White Sea and ashore so discovering the Archangel peninsula. Within a few years, a British colony had become established at Archangel based on trade with 'the land of Cathay' which lay somewhere to the east, its exact location apparently known only to the local natives. They had skirted the White Sea and dragged their flat-bottomed boats, which were sewn together with willow roots, overland to the Kara Sea. Sailing on through the Gulf of Ob, they had reached the estuary of the River Tazz and there, in a town called Mangaze, they had discovered a true Cathay. The town had an annual fair to which all the local merchants travelled via the River Ob and nearby River Yenesei bringing tea, silks and spices. The Archangel men bartered for these fabulous riches and established a regular trade link along the northern coast of Russia. As the port and community of Archangel thrived and grew, imports of tea and spices from Britain gradually put an end to trade with Mangaze. It was not until late in the nineteenth century that any effort was made to reopen the trading route to the east. In 1874 and 1875, an Englishman, Captain Wiggins, and a Swede, Professor Nordenskiold, succeeded in reaching the estuaries of both the Ob and the

Yenesei. Captain Wiggins returned again in the summer of 1876 with the steam yacht *Thames*. He failed to get the wind he needed to sail up the Ob and instead, in worsening autumn storms, ran for the Yenesei. Finally, on 17th October he laid his ship up for the winter on the Arctic Circle, nearly 1200 miles from the river's mouth. The crew were made comfortable in a peasant house on the riverbank; Captain Wiggins made his way back to England overland.

When Seebohm heard that Wiggins was in England and intended to return to his ship in spring, he lost no time in contacting him. His haste was justified, for only five day later, on 1st March 1877, Seebohm and Wiggins started out for the Arctic. As his friend Bowdler Sharpe later said, Seebohm 'was not a man to take long to make up his mind'. The two men travelled the first 2400 miles to Nishni Novgorod train in just ten days, including a three day stop in St Petersburg.

Ahead of them lay a sledge journey of 3000 miles. They fell into the routine of travelling day and night with the bells of the sledge tinkling to drive away wolves, and stopping only to change horses. The whole journey used about 1000 horses, 16 dogs and 40 reindeer. Part of the way ran along the frozen River Volga where Seebohm, as always on the look out for birds, noticed holes in the river bank marking sand martin colonies. They crossed the Ural mountains with surprising ease, 'no more than a succession of hills, the loftiest hardly high enough to be dignified with the name of mountain' according to Seebohm.

A few days later, they were approaching Omsk across the great steppes of western Siberia. 'For nearly a thousand miles, hardly anything was to be seen but an illimitable level expanse of pure white snow. Above us was a canopy of brilliantly blue sky, and alongside of us a line of telegraph poles crossed from one horizon to the other.'

The cost of food and the hire of horses had dropped once they reached the steppe. A ton of wheat, Seebohm discovered, could be bought in the local market for the same cost as a hundredweight in England.

By the time the two men reached Krasnoyarsk, nearly three days journey beyond Tomsk, the snow was melting so fast that they were forced temporarily to take to wheels and to carry the empty sledge along with them. At last, on 5th April, they reached Yeneseisk. Now that the threat of the spring thaw preventing them from reaching the ship had passed, they were able to rest for a few days in a friend's house in the town. Seebohm found a young Jew called Glinski who spoke some German and whom, with great misgiving, he hired as an assistant. 'Glinski was, without exception, one of the greatest thickheads that I have ever met with', concluded Seebohm after teaching him to skin a bird for the first time. But he had to admit that the boy eventually learnt to skin better and quicker than Seebohm himself, and whilst in his employment prepared over a thousand skins.

The thaw had cut up the roads badly by the time the party left Yeneseisk and they needed a team of up to five horses to pull the sledge. In the houses where they stopped, Seebohm found he could buy furs including bear, ermine, grey squirrel, stone-marten, otter, and Arctic fox in its blue or white colour phases. They enquired for sable and the rare black fox skins but these were far more expensive and anyway reserved for the Yeneseisk fur-merchants. They stopped at Turukansk, a village which had acted as host to the famous annual fair when the town of Mangaze was destroyed by the Cossacks in the early seventeenth century. There they visited the monastery and saw the relics of the patron saint of Mangaze which had been saved when fire had swept through the town.

On his arrival on the Yenesei the previous autumn, Captain Wiggins had brought with him a cargo of materials and other tradeable goods. These had been loaded onto the sledges of a local trader named Sideroff who was also preparing to overwinter his ship on the Yenesei, and entrusted to the care of his men led by an over-ambitious character called Schwanenberg. The caravan eventually had to be abandoned amidst petty squabbles between Schwanenberg and local officials. The cargo was now returned to Wiggins in Turukansk; and he, Seebohm and Glinski spent several tedious days selling off ribbons, printed calicos and silks which were, in Seebohm's opinion, 'unsuited to the market . . . and priced at more than double their value in England.'

To add to his irritations, Seebohm found few birds around the village apart from the usual carrion crows and snow buntings. Their host during this time Seebohm describes as an 'old-fashioned Russian official':

> I asked our host to choose a knife or two out of the stock I brought with me for presents, he immediately took six of the best I had, and the day following asked me for a couple more to send to a friend of his at Omsk. He offered me a pair of embroidered boots for six roubles. I accepted the offer. He then said that he had made a mistake, and that he could not sell them, because he had promised to send them to his friend in Omsk. Half an hour afterwards he offered me the same pair for twelve roubles; I gave him the money, and packed them up for fear his friend in Omsk should turn up again, and I might have to buy them the next day for twenty roubles.

No one was sorry to leave Turukansk. Seebohm wondered how so miserable a little place even came to be printed in capital letters on their maps.

The road from Turukansk to Kureika where the *Thames* lay was little used, and its snowy surface had not been trodden hard enough for the big sledge. They spent the remainder of the journey packing and repacking their luggage onto various smaller sledges drawn by horses, dogs or reindeer. Finally the expedition reached the ship on 23rd April 1877.

8.3 *In summer Seebohm had to resort to a slow and uncomfortable native cart called a rosposki for transport*

Seebohm calculated that his share of the seven week journey from London had cost him £87, not counting the purchase of skins.

The Yenesei is the world's fifth longest river, exceeded in length only by the Missouri-Mississippi, Amazon, Nile and Yangtse. Although Kureika lay about 750 miles from its mouth, the river was nearly three miles wide. They found the crew of the *Thames* cheerful and healthy, thanks to limejuice and dried vegetables which had helped prevent scurvy during the winter. The ship was aground on the north shore of the main river bed, close to a small backwater into which Captain Wiggins hoped to move her as soon as the water began to rise. They settled in the farm nearest the ship to await the arrival of spring. Seebohm bought a pair of snow-shoes and was soon shuffling around the forest collecting birds.

Once again, he waited impatiently for the return of breeding birds, and turned his attention to the local Ostiaks, a race of people related to the Samoyedes. One evening an Ostiak man and his son from the nearby chooms came down to Seebohm's cabin for a drink of tea. As he was beginning to wonder what the man wanted, the boy shyly pulled out from under his fur coat a squirrel and a hazel-grouse which his father had shot. Then, not to be outdone, the old man produced a live fox cub from his sleeve. It was sooty black with a white tip to its tail. The Ostiaks were dispatched with instructions to search for more live foxes, and the following day they returned with a vixen and five more cubs. Seebohm reared two of the cubs using a bottle and the fingers of a kid glove in the hope that they might turn out to be the rare black fox. They grew up tame but timid and very fierce with each other; and, to Seebohm's disappointment, they soon developed red coats.

The Ostiaks were hunters, dividing their time between the birds and fur-bearing animals of the forest during winter, and the

8.4 *Seebohm learning how to move around on snow-shoes*

8.5 In spring the Thames *was carried upstream with the ice flow which grounded her on submerged banks*

fish of the Yenesei and its tributaries in summer. They had no reindeer but used sledges drawn by half-famished mongrel dogs. One day when Seebohm had fired into a flock of snow buntings, he watched the Ostiak children catch the injured birds and, later, he saw the birds plucked and hungrily eaten raw.

When at last the thaw began, five weeks after the arrival of Seebohm and Wiggins, things did not go according to plan. The ice melted rapidly on the headwaters of the Yenesei further south, but remained fast at the river's mouth, to the north. At Kureika, the water level rose, the ice broke up and soon the river bed was full of small ice flows rapidly drifting upstream with the current. The creek into which Wiggins had hoped to coax the *Thames* was still high and dry, but the ship, now refloated, was drifted out into the press of ice and carried slowly south. The buffeting ice flow broke off the rudder and, before they could get up sufficient steam to head for a sheltered creek on the south shore of the river, the current had swept the *Thames* aground on a bank. There followed a series of refloatings and groundings until the river level sank by two metres as the estuary cleared, leaving the *Thames* stuck safely in the creek opposite where the crew had wintered.

While the ship was being repaired and refloated, Seebohm returned to his collecting. The migrants were arriving in large numbers and, to his delight, among them were many waders and some thrushes, both groups of birds in which he specialised. He added golden plovers, wood sandpipers, Temminck's stints and terek sandpipers with their peculiar upturned bills to his collection. He found fieldfares preparing to breed, and shot a beautiful and rare Siberian ground thrush. As he explains how he solved one puzzle concerning the distribution of the willow warbler, Seebohm comments:

> It seems too bad to shoot these charming little birds, but as the 'Old Bushman' says, what is hit is history, and what is missed is mystery. My object was to study natural history, and one of the charms of the pursuit is to correct other ornithologists' blunders and to clear up the mysteries that they have left unsolved.

Later he expresses more uncharacteristic regret at having to shoot a little bunting. The eggs of this Arctic species had never been collected before so that proof of the bird's identity was essential. 'It hovered about so close to me', wrote Seebohm, 'that to avoid blowing it to pieces I was obliged to leave the nest and get a sufficient distance away. It seemed a shame to shoot the poor little thing ...'

The first steamer reached Kureika in mid June. It was a paddle-boat from Yeneseisk and it brought the news that Russia had declared war on Turkey, and that England — fortunately perhaps for Seebohm's party — had been prevented from going to the aid of Turkey by a revolution in India. The previous winter Seebohm and Captain Wiggins had arranged for a small schooner to be built for them in Yeneseisk. The next steamer to arrive had the new ship, which Seebohm had christened the *Ibis*, in tow. The engineer, named Boiling, was almost as keen a naturalist as Seebohm.

By the end of June, the migrants were starting to thin out as birds further north began to breed. Seebohm added the eggs of another northern thrush, the redwing, to his collection. He was able to conclude from his observations that the peak bird migration on the Yenesei occurred about a month later than in the valley of the Petchora.

When the river was clear of ice the *Thames* set off downstream under sail. Almost immediately she ran aground and was refloated only by the assistance of her steam winch and cables. During this manoeuvre the wind suddenly changed, driving the ship back onto the bank and into shallow water. All baggage and the supply of wood for the boilers were unloaded into the diminutive *Ibis*, but the ship was stuck fast with the water level falling rapidly. 'Thus ended the career of the *Thames*', wrote Seebohm, 'a melancholy close to a long chapter of accidents and hairbreadth escapes. The ship seemed fated.'

Wiggins decided that the expedition should attempt to return to England in the *Ibis*, and immediately he began to have her construction completed using the rigging from the *Thames*. His crew did not share his enthusiasm for this idea and soon he had a near mutiny on his hands. Captain Wiggins, Seebohm noted privately,

> is an Englishman to the backbone, possessing the two qualities by which an Englishman may almost always be recognised. One of these is an unlimited capacity to commit blunders, and the other is indomitable pluck and energy in surmounting them when made.

On 9th July, leaving half the crew, the dogs and the foxes behind, Wiggins started out with the *Ibis* for Golchika, at the mouth of the Yenesei. Despite a gale, the little boat made good headway downstream. When they reached the point where the schooner belonging to Wiggins's rival Sideroff had spent the winter, they learnt that nearly all the crew had died from scurvy, and that the ship had been destroyed by the pressure of the snow and ice.

When they reached Golchika, Seebohm arranged to spend a day collecting on the tundra with Boiling and Glinski. As usual he was delighted with the birds he found which included Arctic terns, red-throated pipits, phalaropes and long-tailed duck. His only disappointment was that due to what he regarded as 'Captain Wiggins's blunders' he was too late to collect the eggs of the Asiatic golden plover. The birds already had young and he could find no nests.

Seebohm and Boiling had moved from the cramped quarters on the *Ibis* to a steamer which was shortly to head upstream to Yeneseisk. Meanwhile, Wiggins and his old enemy Schwanenberg were arguing over the *Ibis*. Seebohm agreed to sell out his share in the ship so that she could return to Europe, but the two men could not decide whether she should sail to St Petersburg as Schwanenberg wanted, or to the Ob as Wiggins demanded. Since Schwanenberg's crew were still loyal to him, and Wiggins's were not, the final destination was St Petersburg. Meanwhile, Seebohm listened in amazement to the arguments:

> To attempt to cross the Kara Sea in a cockleshell like the *Ibis* was a foolhardy enterprise, and could only succeed by a fluke, but both captains were anxious to risk their lives in the desperate attempt. Ambition and enthusiasm seemed for the moment to have deprived them of common sense.

When the time came to head south again, Seebohm had spent only six days in Golchika, the most northerly village on the Yenesei, and had not been able to find eggs of either the knot, sanderling or curlew sandpiper, but instead he had eggs of three species of willow warblers, the dusky ouzel and the little bunting, all of which had been collected for the first time ever. Two months later, on 12th September, Seebohm arrived safely in Moscow. He went directly to look at the skins from the region of the Urals in the university museum. To his disappointment he found the collection jumbled up and crammed into drawers and cupboards. Meanwhile Captain Schwanenberg and the *Ibis* arrived in Stockholm in early September and reached St Petersburg on 13th December.

Seebohm travelled home to Sheffield in early October, ending a journey which had covered over 15,000 miles. Analysing and recording all the information he had gathered on these trips to Siberia occupied much of the rest of Seebohm's life, and he soon became recognised as an expert on Siberian birds. Pallas had been the pioneer of Siberian ornithology but his book, completed in 1806, had been delayed by the Napoleonic wars, and was not published until 20 years later.

Seebohm found his own observations and skins confirmed Pallas's theory that many Siberian birds, although closely allied with species from Europe or Russia west of the Urals, were completely distinct from them. Seebohm felt that strict adherence to binomial nomenclature (see chapter 1) failed to do justice to these differences between east and west. Instead he proposed, and used himself in all his publications, a system of trinomial nomenclature. Where birds from the European and Asian branches of a species interbred and produced hybrids, Seebohm used a second specific name to denote the intermediate form. This habit earned him much criticism from fellow ornithologists; and today most species distributed across the whole of northern Asia are not considered to be differentiated into separate species at their eastern and western extremes.

Two years after his return from Siberia, Seebohm accepted an invitation from Albert Günther, keeper of zoology in the British Museum, to catalogue the thrushes and warblers in the huge national collection. By the time this was published in 1881, Seebohm had become interested in avifaunas even further east than Siberia. He studied the skins from China donated to the British Museum by the great collector Swinhoe, and he bought several collections of birds from China and Japan. Not content with these, Sebohm paid a Mr P.A. Holst to visit the Bonin and Volcano Islands south of Japan to collect more birds for him.

Seebohm had lost no time reporting his ornithological discoveries in Siberia in papers published mostly in the journal *Ibis*. To these he now added a number of papers about birds of the Far East. He was to publish prolifically for the next decade, producing up to ten papers a year and having a total of over a hundred publications by the time of his death. In addition to papers in *Ibis*, Seebohm produced a series of beautifully illustrated books for which he is now perhaps best remembered. These include a multi-volume *History of British Birds*, two books on his trips to Siberia, another on the birds of Japan, and a quarto-sized monograph on the distribution of waders. This last book, in particular, cost Seebohm dearly in time and money, as it had woodcuts on nearly every page and 21 colour plates. Its reviewers, especially in America criticised its 'idiosyncrasies of classification and nomenclature', but it was generally well liked. The American ornithologist L. Steineger, who was a major critic of Seebohm, also disapproved of his book on Japanese birds. He praised its 'exquisite' woodcuts but added: 'Mr Seebohm's rules of nomenclature, as well as their enforcement and application, are entirely his own, and quite unique.'

Seebohm was an elected Fellow of the Linnean and Zoological Societies and, in 1890, was appointed one of the secretaries of the Royal Geographic Society. In 1892, he proposed to the trustees of the British Museum that he catalogue and arrange their entire egg collection, and add to it his own private collection. The offer was accepted gladly and gave Seebohm the chance he needed to work on bird taxonomy. He spent the next three years, with the help of Emily Sharpe, the daughter of his great friend R. Bowdler Sharpe, sorting through 48,000 eggs.

Seebohm died in 1895, at the age of 63, from 'malignant anaemia' (probably leukemia) which had been affecting him since earlier that year. His last public appearance was at a meeting of the British Ornithologists' Club in October, a month before his death. He told his friends: 'I cannot sleep at night for thinking of the classification of birds. . . . I must try and get well, I have so much still to do.' At the time of his death, Seebohm had been working on a monograph of the thrush family, illustrated throughout with hand-coloured lithographs drawn by J. G. Keulemans which were said to be 'the most beautiful illustrations ever designed by that talented artist'. The manuscript of the book and its illustrations passed to the publishers, and Sharpe, who was also assistant keeper of vertebrates in the British Museum, undertook to finish the work. To his alarm, instead of acting as editor of a completed manuscript, he found he had to write three quarters of the text himself and arrange for the recolouring of the final series of plates. The monograph initially appeared in 13 parts, each with coloured plates and costing one pound and sixteen shillings. Only about 250 copies of each were produced but the book was so well received that it was published in two volumes in 1902.

Seebohm was buried in Hitchin, now at the northern edge of London. It seemed particularly appropriate to his friends that as his coffin was lowered into the grave, a thrush came and sang from a nearby tree.

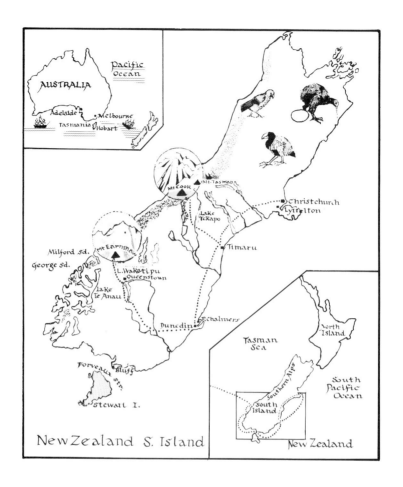

Green's journeys to Austalia and New Zealand, 1882, and the Canadian Selkirks, 1888

CHAPTER NINE
Thousands of Feet Up, Hundreds of Fathoms Down

9.1 Green's unplanned stay in a Melbourne quarantine camp gave him a chance to see some Australian wildlife including wallabies

William Spotswood Green, priest, mountaineer, marine biologist and adventurer, had an unusually varied career for a man of his time. He has been described as 'one of the most remarkable figures in the ranks of Irish science and one of its most stimulating personalities'. Green was born on 10th September 1847 in Youghal, County Cork, on the south coast of Ireland, and brought up an only child. At the age of 12, he was sent to school in Rathmines, a Dublin suburb, and in 1868, after what is reported to have been an 'undistinguished career', he received an MA from Trinity College, Dublin.

The following year Green spent six weeks with his friend J. S. Lyle in the Swiss Alps and, as he said himself, 'I returned home feeling that a whole new world had opened up before me.' Green was a natural climber and straightaway he developed a passion for the stimulating and humbling atmosphere of the high mountains. His only disappointment during the trip occurred when the soles of his only boots burnt off one night beside the fire, preventing him from attempting the peak of Mount Rosa. During the next five years, he made climbing expeditions with his friend to Norway and the Lofoten Islands and, in 1874, he returned to the Alps. Here Lyle was killed in a climbing accident, and in despair Green vowed to keep away from mountains for good. In June the following year he married his cousin Belinda Beatty and in due course had one son and five daughters. Green had entered the Church of Ireland in 1872, and in 1878 he became rector of Carrigaline, a quiet seaside village in his native County Cork.

Since the parish duties were far from demanding, Green soon turned his attention to his second great interest after mountains, the sea, and, more specifically, to fishing which was the chief occupation in the village. 'But', as his obituary expresses it, 'his vocation was altogether too monotonous for him', and he was soon pining for some action. He experimented with a visit to the tropical forests of South America only to be driven home by attacks of fever. In 1879, Green returned once more to the Alps where his health and spirits revived, and his mind became full of plans for mountain climbing.

In 1881, he visited the Jubilee meeting of the British Association which was held at York. Here he happened to see a collection of photographs on display showing the mountains of the so-called Southern Alps on the South Island of New Zealand. Immediately Green made up his mind: 'Mount Cook was a splendid peak, and his conquest well worth the trouble of the long journey'. In addition to this, like Howard Saunders before him (see chapter 7), Green was advised not to spend another cold and damp winter at home. He collected a modest grant from the Royal Irish Academy, and on 12th November he sailed from Plymouth for the Antipodes.

Green took with him two Alpine guides, Ulrich Kaufmann and Emil Boss, from Grindelwald in Switzerland. He equipped the expedition with three large tents, and included one of the new aneroid barometers and a small camera. Unfortunately there was no room for all three men to travel together so they arranged to meet in New Zealand. Green's ship steamed slowly down the coast of Africa, calling at Saint Vincent in the Cape Verde Islands. Here Green took advantage of an afternoon ashore to climb a small mountain overlooking the harbour from which he surveyed the view with vultures wheeling about the rocks below. Sixteen days out from Plymouth they were approaching Cape Town where Green already planned to climb the famous Table Mountain. However, one of the passengers in the crowded third class quarters developed smallpox from which he later died, and the ship was kept in quarantine when it reached port. They were grudgingly allowed to bury the body at sea, and given coal for the ship's bunkers before being directed to land the passengers at Saldanha Bay, five hours north of Cape Town.

They reached the harbour in the evening, steaming in past the guano islands where, even today, so many millions of gannets, cormorants and other seabirds breed that their accumulated droppings or guano is mined and used as a valuable source of phosphate. After three days' delay, they were underway again and heading into the southern Indian Ocean. Seabirds became numerous, following in the ship's wake, coming so close Green could almost reach out and touch them, or sailing overhead. He identified albatrosses, both large and small, and whale-birds or prions. Green was amazed to see the effortless gliding flight of the albatrosses about which he had read so much. Some of these birds, with wingspans of over ten feet, are designed like gliders, with long narrow wings which are rarely flapped. They use the up draughts in the valleys between the waves to gain lift, and they sweep over the sea's surface in a series of graceful rising and falling curves. Using so little energy, the birds are able to cover enormous distances at sea while searching for food.

By the time the ship reached Adelaide in southern Australia, two more people had developed smallpox. Cargo was taken aboard as quickly as possible, and a few passengers were put ashore for the quarantine station on Torrens Island. Then the ship pressed on for Melbourne. Green and the other passengers landed at the quarantine camp on Point Napean in Port Philip Bay, the first time they had left the ship for 45 days. Green's main concern was that he would miss his rendezvous with Boss and Kaufmann and, to his frustration, he saw their boat steaming past the point. He tried to make good use of the delay by exploring the beaches and bush surrounding the camp.

With two friends, he set up camp in the bush where they hunted wallabies and swam from the beach. Around the camp fire one night, though carefully not sitting down for fear of the voracious bull-dog ants, the conversation turned to the controversial theories of the late Mr Darwin. Someone suggested that:

> The progenitor of the kangaroo was an animal with hind quarters of ordinary dimensions like other animals, but whenever he sat down a bull-dog ant gave him a pinch, causing him to make a bound. The constant recurrence of this unenviable phase of his existence through succeeding generations led as a natural consequence to the extraordinary development of the hinder limbs in the present representatives of the race.

Green awakened in the night to find the air 'as still as death' and the bush around the tents alive with game. He crept out hoping to shoot a wallaby, and watched the dawn arrive from the branches of a honeysuckle tree, surrounded by the twitterings, whistlings and chatterings of bell-birds, parakeets, bronze-winged pigeons and wattled honeyeaters.

All too soon they were made to return to the quarantine station by an anxious police sergeant who had been sent to look for them. After a delay of 18 days, Green and the other passengers were allowed to continue their journey. Green was among the first to leave the camp, in a hurry to catch the steamer from Melbourne. A few days later he was on board the *Te Anau* and bound, at last, for New Zealand. Before starting on the 1000 mile sea crossing, the ship called at Hobart on the island of Tasmania where its holds were filled with gum trees split into staves for fencing, cases of greengages, apricots, and pears, and barrels of raspberries for jam whose cloying smell remained with them throughout the voyage.

During the crossing, Green had time to think about the climbing ahead. The mountain chain named the Southern Alps by Captain Cook ran the whole length of the South Island, and several stretches of 100 miles or more had no snow-free passes. The peak of the range was Mount Cook, known as Ao-Rangi in Maori, which had never been climbed before. This, of course, was the main reason for Green's interest in it. In 1862, Dr Julius von Haast and A. D. Dobson had surveyed Mount Cook and reached an altitude of 8000 ft on its southern flank. Eleven years later the Governor of New Zealand had offered official aid to anyone who would climb the mountain but, for the next eight years, no one had taken up the challenge.

The *Te Anau* approached the fjords on the southwest coast of New Zealand in a thick dawn mist. At last Green had his first view of the new country:

> We saw a black rock with a white breaker shooting up its sides. Another moment, and emerald-green foliage showed above it, and as the steamer's head swung round to the northwest, there appeared a high sloping headland all covered with fern vegetation.

Rounding a sea-swept rock encircled by wheeling albatrosses in a furious squall, they entered the calmer waters of Milford Sound. Vertical cliffs rose on either side, and the blasts of wind between them drove the small ship along at great speed. Huge waterfalls plunged deafeningly to the sea and numerous small ones clouded the rocks in a silver veil of mist. Three men in a boat came out to meet them at the head of the fjord, and collected supplies. One of them, who remained aboard, was an explorer who had been trying without success to find a route over the rugged mountain range from Milford Sound to Lake Wakatipu and the inhabited eastern slopes.

The weather had improved enough for the captain to take the *Te Anau* into George Sound.

> We steamed through the most lovely corridor of rich forest scenery, rising tier above tier to the highest limits of the vegetation. On and on we went, past an islet covered with fine trees draped with lichens, the whole reflected gem-like in the still water... We slipped past a point and entered a basin in which we were quite shut in from the view... Immediately before us the foaming fall plunged into the sound, filling the air with its roar. For a moment we felt as if we were at the bottom of a deep well, so small was the patch of sky overhead, the walls of the forest all around rising rapidly for 3000 or 4000 feet.

When Green awoke the next morning, the ship was steaming through Foveaux Strait, between Stewart Island and the mainland, with a train of albatrosses and black and white petrels called cape pigeons in its wake.

He went ashore briefly at Bluff where two telegrams were delivered to him, one from Boss and the other from the Director of the Geological Survey welcoming his party to New Zealand. Soon after the *Te Anau* reached Port Chalmers the following morning, Emil Boss stepped into Green's cabin. Green and his friends spent a few hours in Dunedin. He found it 'A great city ... with its splendid streets and terraced hills, bright with charming gardens and pretty villas, and then to think that forty years ago Dunedin did not exist!' Soon they were aboard the *Te Anau* again and heading for Christchurch over a glassy sea with seabirds skimming to and fro around the ship. The coastline, in contrast to the fjords of the west side, had green valleys, snug coves and picturesque volcanic rocks. Far away on the horizon was a serrated line of mountains, topped with snowy peaks.

When they reached Port Lyttelton, Green hurried ahead by train to Christchurch. Here he visited Dr von Haast in the Canterbury Museum, a man he felt he already knew well from reading and rereading his book about the Southern Alps. Together they examined the museum's large collection of moa skeletons. These gigantic ostrich-like birds stood up to ten feet high; endemic to New Zealand, once they were widespread throughout the country. The last species of moa is believed to have become extinct about a hundred years before Green's visit. Another endemic species extremely rare in Green's time was the takahe, a brightly coloured rail about the size of a small turkey. The bird was known only from fossil remains up to 1849 when the first skins were collected. There were a few reports of the species in the next decades, but it was thought to have become extinct by 1898. A deliberate search made 50 years later revealed the first live takahes ever seen, near Lake Te Anau. Today, a few hundred takahes live in the Murchison Mountains above the lake.

Next, the two men examined von Haast's map of the Southern Alps, and photographs of Mount Cook. The following day, Green and his two friends travelled south by train to Timaru. Here they stocked up on supplies by buying flour, oatmeal, biscuits and some tea, rice, cheese and tinned sardines. The next train took them up 1000 ft to Albury, a small collection of buildings at the end of the railway track, set in rolling downs. Finally they obtained three horses and an ancient wagon for their journey into the mountains, and a few days later on 9th February the expedition was ready to start.

They crossed undulating pasture, occasionally startling a New Zealand harrier from a sheep carcass. They stayed a night on an island in Lake Tekapo at the foot of the mountain range. While walking beside the lake before supper, Kaufmann saw their first weka, a rail-like bird found only in New Zealand, scuttling through the undergrowth.

From Lake Tekapo they followed a bullock track to the Braemar sheep station in the Tasman River valley, and here they had their first view of Mount Cook 'towering over all, blocking up the vista to the right with his pyramid of rock and ice, and forming one of the grandest scenes of the southern hemisphere'. By midday, they had reached Mount Cook sheep station where they hoped the proprietor, Mr Burnett, would be able to help them over the Tasman. As Green explains,

> the New Zealand rivers are so swift and erratic in their courses that fords are dangerous and bridges difficult to construct. Once the rivers leave the mountains there is nothing to keep them to one channel, as the plains, being composed of loose boulders and sand, are easily eaten away by the swift streams. . . . A river bed is therefore a broad sheet of gravel through which a number of small streams wander and change day by day.

The Tasman River bed was two miles wide with fast-flowing channels and dangerous quicksands. The party disturbed black-billed gulls from the marshy flats nearby. The crossing of the Tasman River was accomplished with no more than a broken harness.

> A few miles above where we crossed, the valley forked, and the terminal morain of the great Tasman glacier was now distinctly visible, blocking up the valley to the right, while great piles of debris brought down by the Mueller and Hooker glaciers filled the branch to the left. Right in the middle of the picture, between these two branch valleys, the Mount Cook range rose pile upon pile, its glittering peak crowning all.

A storm was brewing in the Hooker valley and rain began falling at sunset as the party made its way across the gravelly valley of the Tasman River. They spent the night in a shearing shed at Birch Hill with the station shepherd, George Southerland. All night the hail and rain beat upon the iron roof while thunder crashed and blue lightning lit the hut. Green and his friends started out with Southerland early next morning, skirting a sedge-clogged lake under the rock face where they disturbed a flock of paradise ducks and rails called swamp hens.

The roar of the meltwater torrent from the Hooker valley grew gradually louder as they approached it. On the bank they abandoned the wagon, and loaded their baggage and a sheep carcass onto the horses. Before Southerland left with the wagon he helped them across the numerous icy streams of the Hooker River and directed them on to the spur of Mount Cook. One of the horses had chosen to roll in the river so the valuable supply of flour and Green's change of clothes had to be dried out in camp that night. The wekas stood around watching with comical curiosity. Green killed one and found it too oily to cook but good for greasing the boots instead.

Nearby the men discovered a pool full of ducks. Some were blue ducks, about the size of a European wigeon, which never flew if they could avoid it. The rest were the goose-like paradise ducks which had harsh, plaintive cries and heavy flight. Southerland had told them that usually a flock of 2000 or so sheep were driven over the Hooker valley on the glacier to graze all summer long on the grass in the foothills of Mount Cook. In 1882 an ice-bridge on the glacier had broken and so now the sheep could no longer come across. Green, Boss and Kaufmann decided to remedy this by building a bridge over the Hooker above their camp using tree trunks and ropes.

The next day they crossed their bridge and walked up onto the moraine at the foot of the Great Tasman glacier. They agreed to move their camp further up the valley. Two keas, or Mount Cook parrots as Green calls them, arrived while they were caching their equipment near the new camp site. These birds are a dull green with scaley markings on their upper parts but have vivid red and yellow underwings. They are infamous scavengers and often very destructive in their search for food. Green seems to have known this for he piled large stones onto the cache before leaving and wrote: 'I doubt if they would have eaten our [ice] axes, they might have been tempted to try such delicacies as the cartridges and gelatine plates!'

When they returned to their camp, they

found Southerland had ridden up to see them, bringing the news that the wagon had been swept away while crossing the Tasman although the horses were saved. For the trek up to the higher camp their equipment was reduced to about 130 lb and to this was added 100 lb of food. The rest was packed securely into the remaining tent and left to the inquisitive vigilance of the wekas which had become as tame as hens. Green knew that von Haast had had trouble with rats in the mountains and so was surprised to find none on Mount Cook. He assumed the Norway rats, which systematically wiped out the native Maori rats, had advanced into the high mountain valleys and, finding they could go no further and with too little to eat, had retreated again.

The three men moved up to the new camp site and set up the tent when a terrific storm broke over the valley. The next day was passed in the leaking tent with the temperature little above freezing point. They moved on in stages up the glacier, carrying all they needed to successively higher camps. The fifth camp was made at 4000 ft where a small glacier joined the Tasman; this Green named Ball Glacier after John Ball, the first President of the English Alpine Club. Even at this altitude there were surprises. Green saw a single 'large sea-gull' winging its way up the glacier, and by its speckled plumage he decided it was a young bird migrating to the Pacific. He gives no further details but the bird seems most likely to have been a young black-billed gull.

Provisions, supplemented by ducks and keas, were brought up to the top camp and, on 1st March, after a few unsuccessful trials, they were ready for the assault on the peak of Mount Cook via a spur of rock north of the Hocksetter Glacier. The ridge they followed was isolated by the glaciers and had no plants growing on it although they saw one small tailless wren. By evening, they had reached the top of the ridge at about 7500 ft, and there they camped for the night.

The next morning great banks of cloud were settling in the valley with the mountain peaks projecting like islands in a silver sea. The day's climb became a race against time to the 'endless booming and crashing of avalanches'. After travelling from the other side of the world to stand on the top of Mount Cook, they refused to give up despite the worsening weather. The blasts of ice-laden wind were so powerful that the men had to crouch and cling to their ice axes embedded in the snow to prevent being swept away. At last, with less than an hour of daylight left, they stood on the summit:

> Our first glance was, of course, down the great precipice beneath us towards the Tasman Glacier — the precipice up which we had gazed so often — but the dark grey masses of vapour swirling round the ice-crags shut out all distant view. A look backwards, down into the dark, cloud-filled abyss out of which we had climbed, was enough to make us shudder, it looked fathomless; and this white icy ridge on which we stood, with torn mists driving over it before the fierce nor'wester, seemed the only solid thing in the midst of the chaos.

In fact, some confusion surrounds Green's claim to have reached the top of Mount Cook. Certainly he thought the ridge upon which he and his companions stood was the mountain's highest peak. He estimated the altitude to be between 12,300 and 12,500 feet, which agreed with the height of 12,349 feet obtained trigonometrically by the Westland Survey Department. This is the height of Mount Cook as marked on modern maps. Neither Green's friends nor his biographers doubted the authenticity of his claim to the summit of Mount Cook. However, an entry in *Everyman's Encyclopedia* records that Green only 'explored' the mountain in 1882, and that the first ascent was made by the New Zealanders, Fyfe, Clarke and Graham, in 1894.

With darkness falling fast, the weather deteriorating every minute and the fatigue of their long climb overtaking them, this must have been the last thing on the minds of the three men as they stood at the top of Mount Cook. In Green's words: 'The great problem we had hitherto longed to solve was: Could we get up Mount Cook? That question was now settled. A more anxious problem yet awaited solution: Could we get down?'

The descent was hurried and in places, Green admits, hair-raising. Fortunately the moon was full and its light diffused through the clouds. At last, they reached a sheltered ledge, a few feet wide and still at an altitude of over 11,000 ft, on which to pass what was left of the night. A warm northwesterly wind began blowing, and all around them the snow started melting and running down the rocks. At least four times an hour they heard avalanches in the distance which rumbled and shook the rock under their feet. At 4.30 a.m., after eight hours of keeping each other awake, the dawn began to show on the cliffs of Mount Tasman and an hour later they were on the move again. When they reached the Linda Glacier they found that their ice-bridge was gone, and instead they had to crawl over a crevass on a very unsafe bridge. Further on they had to detour the ice from an avalanche they had heard the previous night which now blocked their path. Had they decided to continue on down the mountain in the dark, they calculated, this avalanche would have buried them. At last, at 7.30 that evening, they reached camp after an expedition lasting 62 hours.

The next day was spent sleeping, eating, tending to swollen hands and writing up diaries. Among the alpine plants Green had collected was one with a white flower which grew in velvet-like clumps well above 6,000 ft. Later, when this was shown to Sir Joseph Hooker in Kew, he pronounced it a new species and named it after Green. The three men set out once more down the Tasman Glacier, packing up each camp as they came to it. Crossing the Hooker River swollen by the brief thaw proved very tricky and, from his own account, Green nearly drowned. By the evening of the next day, they had been reunited with Southerland at Birch Hill, and forded the Tasman River on their way to Mount Cook station. Here, as they were sitting down to a meal of mutton, the door opened and in walked an elderly gentleman in a wideawake hat, black coat and neat boots, carrying nothing but an umbrella. He was an American who was lecturing in Dunedin on fossils and 'the history of the earth', and was invited to share the mutton. Dessert was taken in Mrs Burnett's garden where they ate gooseberries and red currants from the bushes.

At the Braemar station, where they had their last view of Mount Cook, the expedition was met by the gentlemen of the press. 'What sort of mountain is Mount Cook?' they demanded to know of Kaufmann. His reply, to Green's delight: 'Mount Cook is still there, and if you want to know what it is like, you can go and see.' By 13th March Green and his men were back at Timaru on the coast, and the next day they travelled to Christchurch in an unsuccessful attempt to see von Haast.

As the steamer for Melbourne did not leave for another week, Green accepted an invitation to visit the district of Otago and climb Mount Earnslaw. The three men took a train up from the coast to Lake Wakatipu which lay shrouded in night and surrounded by echoing mountains. The lake, Green learnt, was extremely deep; its surface was 1100 ft above sea level and its bottom 430 ft below it. It had been stocked with trout from a Tasmanian colony, founded on fish brought from the River Wye in England. One trout Green saw soon after it was caught weighed over 15 lb. The men took a paddle-steamer up the lake to Queenstown, and the next morning they saw the glaciers and twin peaks of Mount Earnslaw shining through the mist. They rode to the foot of the mountain and climbed as far as they had time for towards its summit. A few days later, with no time to spare, they boarded the steamer for Melbourne.

In less than two months, Green was back in Plymouth. 'Travelling on without delay', he reports, 'I reached home just in time to take morning service in my own church.' One had almost forgotten he was a priest. 'A week later I resumed all my old home life and occupations and the past six months ... seemed like a pleasant dream', a familiar feeling to anyone returning home from a stay abroad.

In June of 1882, Green applied to the

Secretary of the Royal Irish Academy and was given permission to read a paper entitled 'Recent explorations in the southern Alps of New Zealand' to the Academy's next meeting, thus earning himself considerable respect for his expedition. The following year, his account appeared as a book *The High Alps of New Zealand* published in London by Macmillan. In the same year he read a paper on the southern Alps to the prestigious Royal Geographic Society.

With this reputation established, Green seemed a good choice for organiser of the Royal Irish Academy's deep-sea dredging operation which started in 1885. In response to a new and growing interest in deep water life around Europe, the Academy provided a grant for the investigation of the fauna within the hundred fathom line off southwest Ireland. Green organised a series of expeditions for them during the next three years.

The first was a five day trip in August, headed by Professor A. C. Haddon of the Irish Royal College of Science, with ten biologists aboard the paddle-steamer *Lord Bandon*. One of those on this trip was Samuel Haughton, who had helped Leopold McClintock with his geological specimens from the Arctic (see chapter 4). The expedition dredged down to 120 fathoms at one point northwest of the pinnacle stack of Great Skellig off the coast of Co. Kerry and retrieved many interesting marine species. The following January, Green was elected a Fellow of the Royal Geographic Society.

The dredging expedition had been such a success that in May 1886 Green was asked to organise another one, again led by Haddon. They made an unsuccessful attempt to dredge at 1100 fathoms but managed to collect samples by dredging and trawling in a few hundred fathoms of water. Green developed

9.2 The dredging crew of the paddle-steamer Lord Bandon *in 1885:*
(left to right) Professor A.C. Haddon, Rev. Samuel Haughton, Rev. W.S. Green, and C.B. Ball
(Courtesy of the Dundalgon Press, Dundalk)

9.3 Among the specimens raised from a depth of 750 fathoms by the dredging expedition off Cobh were starfish and sea urchins

the use of wire hawsers for deep-water trawling, and described the equipment used in a joint report of the expedition with Haddon, published in the *Proceedings of the Royal Irish Academy*. For Green there was a special thrill to deep-sea dredging especially 'when hauling some 100 fathoms of trawl warp, my eyes first behold a rare creature, whose form I had long hoped to find.' Among the numerous animals they collected in 1886 was a new species of pink anemone coated in sand which was named in honour of Green.

Green became increasingly involved with the Irish fishing industry in 1887 when he was asked to write a report for the Royal Dublin Society on how it could be promoted. In December he approached the Royal Irish Academy for a grant towards the cost of a third dredging expedition to collect specimens from further offshore where the water was well over a mile deep. The ship used was the *Lord Bandon*, renamed the *Flying Falcon* (not the *Flying Fox* as suggested both in Green's obituary and in Went's account of his life and work). The party of seven biologists included one of Ireland's best loved naturalists, Robert Lloyd Praeger. One trawl at 1270 fathoms took an hour to raise to the surface. Another at 750 fathoms collected

> great sea-slugs, red purple and green; beautiful corals, numerous sea-urchins with long slender spines; a great variety of starfishes of many shapes and of all colours, including one like raw beefsteak, which belonged to a new genus; strange fishes, and many other forms of life. This and previous hauls provided a fine display of the nature of the fauna which lives in complete darkness on the sea-floor a mile or more below the surface.

Praeger immediately began to wonder why the animals, apparently living in perpetual darkness, bothered with such brilliant colours.

That evening the weather became very rough. The *Flying Falcon*'s starboard paddle-box and galley were smashed in as she steamed for shelter. Praeger recalls the night that followed:

> I remember that when at last grey dawn gladdened our sleepless eyes, Green's face, ghastly white under a sou'-wester, appeared

at the cabin door. He looked round at the wreckage strewn on the floor, but all he said was, 'It has just occurred to me that if a hippopotamus had been shown his photograph before he was created, he would have been able to suggest some important improvements!' The trivial incident was quite characteristic of the man. He bore any hardship with a laugh, and no emergency disturbed his whimsical good humour.

In the long and twisting entrance to Cork harbour the next morning in thick fog they narrowly avoided collision with a White Star liner. This expedition was recorded in photographs taken by R.J. Welch, glass negatives of which apparently are now in the Royal Irish Academy in Dublin.

Despite his involvement with the sea, Green had not forgotten his great love for mountains. In June 1888 he set out for Canada, with his cousin the Rev. Henry Swanzy, to explore the Selkirk Range in the British Columbia Rockies. The region had been opened up a couple of years earlier by the completion of the Canadian Pacific Railway, and the two men were commissioned by the Canadian Government to survey the peaks and glaciers of the Selkirks. Swanzy had visited the Canadian Rockies before, on an excursion at the time of the British Association meeting in 1884.

Green and his cousin arrived in New York and travelled by train to Ottawa, and thence for four days on to the Rockies. (Went's account is again at fault here in stating 'they proceeded from New York to Montreal by ship'.) The Selkirks were a wetter range than the rest of the Rockies and their lower slopes were covered in luxuriant and nearly impenetrable forest which is why the area was so

9.4 Caribou still live in the Selkirk mountains but in 1984 they were added to the U.S. endangered species list. In the year of Green's trip to the Rockies, Theodore Roosevelt, President of the United States, visited the Idaho Selkirks to hunt caribou

poorly known. The forest included several trees rare in the main Rockies such as western white cedars and western hemlocks. The wildlife, on the other hand, tended to be less varied with no bighorn sheep and just a few wild goats. There were both cinnamon and silvertip varieties of grizzly bear, and considerable herds of caribou. The birds included ducks and geese on the high lakes, three kinds of grouse, and numerous white-tailed eagles and ospreys, called fish hawks, on the Columbia River bordering the range.

The two men and their companion, Richard Barrington, with 'Jeff' the dog and several pack-mules, surveyed the region around their base at Glacier House, and made a modest 13 hour ascent of Mount Sir Donald. The alpine flowers were at their best with yellow lilies, large purple daisies and anemones. They saw grouse and large hoary marmots, also called whistlers, which picked bunches of the flowers to line their burrows. Green made an excursion to Vancouver on behalf of the Royal Dublin Society to visit a salmon cannery supplied by native Indians in canoes and manned by 'Chinamen'.

In early August, with Jeff and the mules, the party set out to scale the 11,500 ft Mount Bonney. Green was impressed by the beauty of the trail which led them through carpets of yellow mountain avens and a

9.5 *Green's cousin Henry Swanzy descends a slope on the Asulkan Glacier below Mount Macdonald to hunt a goat (Courtesy of the Bodleian Library, Oxford)*

> meadow of *Veratum viride* as high as our waists. Here again a bear was making tracks ahead of us. The broad leaves were so filled with dew that walking through was as wetting as if we were in the river, we gladly followed the path he made; the stems being crushed down and the dew shaken from the leaves in his wake.

By eleven that night, having climbed the mountain, they were 'round a blazing fire, sipping chocolate and picking the bones of a marmot.'

Next Green and his friends moved up the valley of the Illecellewaet, or Roaringwater River, on horseback with a guide from the Selkirk Mining Company. They passed through 'an alpine garden' which even Green had to admit was better than any he had seen before, and on up Beaver Creek to the forest at the foot of Mount Macdonald. Here the going became rough because of numerous fallen trees, but eventually they emerged from the tree line to survey the Asulkan Pass. Typically, Green had with him a rope that

> had seen some vicissitudes of fortune in my company. Its first good work was to save the lives of some of our party in a bad slip near the summit of the Balmhorn in the Bernese Oberland. It was next used as the mizen topping-lift of a 15-ton yawl. It was my tent-rope in the New Zealand Alps. It was a bridle used on a deep-sea trawl that went down to 1000 fathoms beneath the surface of the Atlantic. It trained a colt. Now it was in our diamond hitch, and I regret to say that its old age was disgraced by its being

used for cording one of my boxes on the voyage home.

In early September the expedition headed east, by lake steamer and train, to Banff and so eventually to Ireland. Green's account of the expedition, *Among the Selkirk Glaciers*, was published by Macmillan in 1880, and the results of his surveys appeared as a map in the *Proceedings of the Royal Geographic Society* in 1889.

The year after his return from Canada, Green organised his fourth deep-sea dredging expedition, this time at the suggestion of Dr Albert Günther from the British Museum. The expedition included Robert Ussher, one of the founders of Irish ornithology, who 'was asked to transfer his affections from birds' eggs to sea eggs' and take charge of drying and packing echinoderm eggs. The trawls were made down to over 1000 fathoms from near Cork harbour right out to west of the Fastnet Rock. A number of species new to the Irish fauna were collected and lengthy scientific reports, including one by Günther, were appended to Green's account of the expedition.

Praeger quotes the author Stephen Gwynn (also Mary Kingsley's biographer) as saying that the only ecclesiastical post that Green would really have liked would have been that of chaplain on board a pirate ship! In January 1890, Green left the ministry to become Inspector of Fisheries in Dublin. The Government agreed to allow him to conduct a survey of fishing grounds between north Donegal and west Kerry for the Royal Dublin Society. This took up most of his time for the next two years, and involved several expeditions up and down the west coast collecting information and interviewing fishermen.

In 1892 Green was appointed to the newly formed Congested Districts Board to advise it on how fisheries could be improved in the poorer western districts of Ireland. One of his achievements, for example, was to get the Board to equip the Aran islanders of Galway Bay with large boats and suitable gear 'to catch mackerel in greater quantities when they came inshore at harvest-time, ... also to fish for them in May from outside the islands — a new venture which proved highly successful.'

Green and his friends had long been fascinated by the idea of landing on the tiny, steeply conical island of Rockall, 250 miles northwest of the Donegal coast. In 1896 Green consulted Professor Haddon, R.L. Praeger and the ornithologists R. M. Barrington and J.A. Harvie-Brown, who had accompanied Seebohm to Siberia (see chapter 8). They decided to ask the Royal Irish Academy, of which Green had been elected a member the previous year, for a grant towards the cost of a Rockall expedition. The Congested Districts Board agreed to charter their steamer *Granuaile* to the expedition, and in early June they set out for the rock, accompanied by Green's son, Charles, who was to be the photographer.

To the delight of the ornithologists on board, several great shearwaters, visitors from the Falkland Islands or Tristan da Cunha in the south Atlantic, were sighted during the voyage. Green describes Rockall thus:

> The little dark, lonely rock, in shape much resembling a haycock, seemed but a speck in the stormy sea. Girt in seething foam the great swells, as they struck it, sent up single jets of spray, sometimes higher than its summit, and, as they met around its lee and dashed together, completely enclosed it in their embrace.

The rough sea prevented a landing, and the expedition was forced to return to Killybegs for more coal. They returned to Rockall again five days later and circumnavigated the island. Green tried to snatch a specimen of seaweed from the rock but had to be content with retrieving some shearwaters, a tern and an immature kittiwake shot by Harvie-Brown and Barrington. A number of dredging hauls were made around the rock and its extensive banks.

At the end of his account of the Rockall trip Praeger echoed what must surely have been the feelings of most, if not all, of those present by writing: 'I shall never see again that desolate and lonely rock, but I feel a deep affection for it, because it brought me a fortnight's intimate fellowship with some of the best companions it has been my privilege to know.'

Green published a report of his visit to Rockall in the *Transactions of the Royal Irish Academy*, and illustrated it with several of his own watercolours (see C17) and a series of rather shaky photographs taken by Charles Green.

For the next 18 years, Green occupied himself with work for the Fisheries Office, the Royal Dublin Society and the Department of Agriculture and Technical Instruction. Praeger describes him thus:

> I used to think of him in his room in Kildare Place as a kind of shorn Samson. Sometimes I met him going home with a bundle of papers over which he must spend the evening . . . Slow routine and a red-tapish atmosphere must have chafed that ardent spirit, but the conviction that he was helping the fishing people carried him on, and he was an almost over-conscientious officer in his imprisonment.

In 1914, Green retired and moved back to his beloved west coast to live the rest of his life in West Cove, County Kerry. 'He had satisfied', Praeger says, 'the two chief desires of his life: he had explored the high mountain-snows of two continents, and also the depths of the ocean.' Green died on 22nd April 1919, at the age of 72, and was buried at Sneem, County Kerry, within sight and sound of the Atlantic swell.

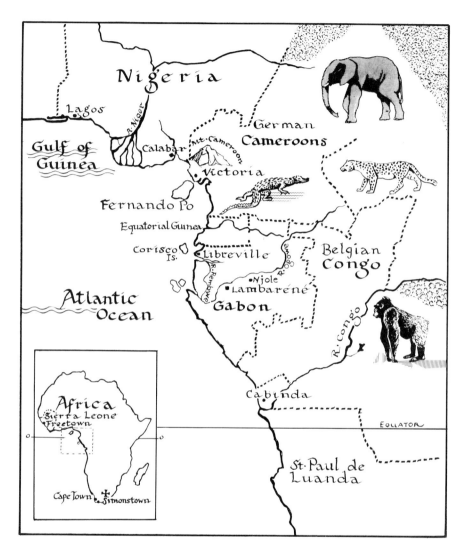

West Africa, showing places visited by Mary Kingsley, 1893–5

CHAPTER TEN
Fishes and Fetish

10.1 Mary Kingsley in 1896 or 1897, signed 'the melancholy picture of one who tried to be just to all parties.'

'Mary Kingsley lived in a man's world where the intellectual horizons of women were supposed to be, and usually were, bounded by spheres of household management, fashion, mothercraft, religious piety, and improving kinds of literature.' So wrote Dr J. E. Flint in his introduction to the third edition of Kingsley's *West African Studies* which was published in 1964, 64 years after her death. Despite these quite genuine and severe restrictions, Mary Kingsley made two long journeys through equatorial West Africa, visiting parts of present-day Cameroon and Gabon where no white person had been seen before.

Mary Kingsley was not alone in Victorian female society in her ambitions to travel to the remotest places. Among her contemporaries were Miss North, Miss Gordon Cumming and Miss Bird who had earned for themselves considerable reputations for their various journeys in North and South America, and Ceylon. What sets Mary Kingsley apart is the way in which she 'stepped so abruptly from conventional circumstances into the unknown'; and that she travelled not without purpose but with the specific intention of collecting zoological specimens and of studying anthropology, particularly native fetish customs.

To understand Mary Kingsley and her role as one of the nineteenth century's better-known travelling naturalists, one has to begin in London, where she was born on 13th October 1862. (Curiously enough, in this and the following year, all the naturalists covered in these chapters were alive simultaneously.) Her father, George, came from a family of writers. His elder brother, Charles Kingsley, had written several popular books including the political satire, *The Water Babies,* and an adventure story called *Ravenhoe!* George was also a writer, among many other things, but more significantly he was an absolutely compulsive traveller, quite 'incapable of staying at home' according to Stephen Gwynn, Mary's main biographer. He left his invalid wife, his son Charles and daughter Mary in a small house on Highgate Hill in north London while he travelled with one expedition after another literally all over the world. Mary grew up to the duty of nursing her mother, and of taking care of her brother.

Her father's rare and brief visits home gave Mary tantalising glimpses of the world outside north London. His detailed letters and his descriptions to her of Egypt, the South Seas, the western frontiers of North America, Newfoundland, South Africa, Japan and the Antipodes helped her to build up a vivid imagination. Mary received no formal education whatsoever during her youth, which seems remarkable, even by late Victorian standards. She was encouraged, however, to learn German but only because she could then assist her father's studies by doing translations from books and journals. In his absence, she gleaned a quantity of miscellaneous information from George Kingsley's extensive library. So dependent upon the library was she that when her father threatened to lend a particular volume on solar physics to a neighbour, not knowing that Mary was in the middle of reading it, she quickly hid it away in the shed so it could not be found and sent out of the house. In this as in much of the rest of her youth Mary behaved at times more like a tomboy than the developing Victorian gentlewoman.

Her childhood escapades are related with much humour in an autobiographical magazine article and in many of her private letters, and are quoted extensively by her biographers. She does not seem to have had any particular interest in natural history at this stage in her life, and she kept few pets apart from two game-cocks. These did almost nothing but crow loudly and continually escape from their run, particularly, it seemed to her, when her father was at home. 'There is something fine in a game-cock's crow,' she wrote, 'it is stirring; it used to produce that effect on my father considerably, and I might just as well have crowed those crows myself, for I was held accountable for them.' This particular hobby was brought to an end when her favourite cock finally disgraced himself by attacking the postman and driving him into the cellar.

Throughout her childhood and youth, Mary was constantly at the call of her sick mother, and had no independence at all. On one occasion, when Mary had managed to escape on holiday with the family of her friend Violet Paget, she was summoned home by telegram. In 1879, her father gave up his travels and moved the family to Bexleyheath in south London in the hope of improving his wife's health and his own rheumatism. Four years later, Mary's brother Charles went to Cambridge to read law and the Kingsley family moved with him to a house on Parker's Piece, just outside the town centre.

Mary was 21, a 'slim, pale girl with her straight fair hair, honest eyes and insatiable interest in things of the mind'. Although she was not shy, we are told 'hardly anyone was able to improve on the acquaintance, she was so constantly shut up at home'. However, life in Cambridge did bring her into contact with several people who were to be important friends. One was Violet Paget and another Dr Guillemard, a valuable adviser and sympathiser. But the Cambridge years were especially difficult for Mary who was struggling with her ailing mother. In February 1892, George Kingsley quite unexpectedly died in his sleep and he was followed just six weeks later by his wife.

Mary was desolate at the loss of both parents around whom her life had revolved. Her father, whom she admitted was selfish and quick tempered at times, had been the focus of her love and admiration. In her own words, there was

> something wonderfully attractive even in the appearance of this lithe, square-shouldered man. His strong, mobile face was sunburnt and weather-beaten like the face of a sailor; his fearless brilliant grey eyes looked right into the hearts of those who spoke with him; his whole form was alert and instinct with the warm, passionate spirit of life; and his conversation ranging easily through every subject from philosophy to fishing, full of dry humour and flashing with brilliant wit and trenchant repartee, had a charm which was absolutely irresistible.

After his death, Mary continued in her role of amanuensis, deciphering and publishing a collection of George Kingsley's letters and travel sketches. These appeared in 1899 as a book entitled *Notes on Sport and Travel*.

Mary's brother Charles departed for the Far East within six months of his parents' death. He left Mary with a small personal income of £500 a year, and, for the first time in her 30 years, with responsibilities to no one but herself. Predictably, since there was nothing to keep her in England, she chose to travel. Influenced by her father, she decided to complete his anthropological work by visiting West Africa, one part of the world to which he had not been.

Her choice seems to have been unaffected by the reputation currently enjoyed by equatorial West Africa, the so-called 'White Man's Grave', or by the gloomy prognostications of her friends. The training which her father, quite unintentionally, had given her, strengthened her resolve. The various bits of research which Mary had undertaken at his instruction, covering such subjects as native sacrificial rights, had increased her longing for knowledge. The specimens her father had shot or collected, and assembled in the Kingsley houses, served as a constant reminder to her.

Most contemporary female travellers had graduated to expeditions in little-known countries from holidays at fashionable European resorts. Mary, by contrast, had never been abroad before in her life, apart from a brief stay in Paris and a short holiday in the Canary Islands immediately after her parents' deaths. She was, when she set out for what was then one of the least-known and most dangerous parts of Africa, far from being a seasoned or experienced traveller.

Mary decided that zoology should be the main excuse for her journey. It was certainly one which those around her could more readily understand than a desire to complete her father's notes on fetishism. She was encouraged and instructed on collecting specimens both by her Cambridge friend Dr Guillemard, and by Dr Albert Günther of the British Museum in London. She planned to collect specimens of fish, insects and plants as she had scruples about killing anything larger. 'I am habitually kind to animals, and besides I do not think it is ladylike to go shooting things with a gun', she said.

Although she never did carry firearms, Mary confessed to arming herself with a revolver, a native knife and a small dagger. Despite the difficulties, common to all travelling naturalists and even the modern expedition zoologists, of preparing and preserving specimens whilst constantly on the move, Mary was to succeed in bringing home a better collection than either Guillemard or Günther had expected of her. What had started out as a respectable excuse for travel, became a genuine interest and purpose for her trips.

Her journeys were financed from her own pocket. In order to supplement these funds, and to provide herself with a readily apparent and acceptable reason for exploring remote areas, Mary decided to travel as a trader. She carried a stock of cottons and beads which she exchanged for ivory, palm oil and rubber with the natives along her way, and later sold at a profit. It was not, however, to be an easy profit for the rubber melted in the heat, and the ivory stank and attracted flies.

Mary Kingsley left England for Africa in August 1893, not on a passenger liner but on a small cargo boat called the *Lagos*. As she wrote later in her book *West African Studies*: 'You must go on a steamer that has her saloon aft on your first trip out or you will never understand West Africa.' Captain Murray of the *Lagos* was the first to give Mary the kind of unbiased practical advice she needed. She was also educated in the horrors awaiting her in West Africa by her fellow passengers who were mostly government officials and traders. For some reason, they thought this 'rather gaunt young woman dressed in black' was from the World Women's Temperance Association. Once they learnt that she was actually (in her own words) 'only a beetle and fetish hunter' relations, no doubt, improved.

Her father's tales had prepared her for the first sight of Freetown harbour in Sierra Leone. She knew many details of its history, although mostly from the point of view of its numerous pirates and privateers. She evidently disapproved of its late nineteenth-century sophisticated luxury and formed the opinion that 'Sierra Leone appears at its best when seen from the sea, particularly when you are leaving the harbour homeward bound; and that here its charms, artistic, moral, and residential, end.'

We know relatively little about Mary's first visit to Africa, which lasted about six months, and took her south to Calabar (southeast Nigeria), then to St Paul de Loanda (now Luanda) in Portuguese territory (Angola). On her way north, still using the uncomfortable and unpredictable coastal steamer service, she visited the Belgian Congo, French Gabon (or Gaboon as Mary sometimes called it), and the German Cameroons (modern Cameroon). One of the few episodes she relates from this trip occurred whilst she was staying in a 'simple African home' one night. The peace was shattered by an invasion of voracious driver ants.

> The family, accompanied by rats, cockroaches, snakes, scorpions, centipedes, and huge spiders, animated by one desire to get out of the visitors' way, fell helter-skelter into the street. . . . The mother and father of the family, when they recovered from this unwonted burst of activity, showed such a lively concern, and such unmistakable signs of anguish at having left something behind them in the hut, that I thought it must be the baby. Although not a family man myself, the idea of that innocent infant perishing in such an appalling manner roused me to action, and I joined the frenzied group, crying 'Where him live?' 'In him far corner for floor!' shrieked the distracted parents, and into the hut I charged. Too true! There in the corner lay the poor little thing, a mere inert black mass, with hundreds of cruel Drivers already swarming upon it. To seize it and give it to the distracted mother was, as the reporter would say, 'the work of an instant'. She gave a cry of joy, and dropped it instantly into a water barrel, where the husband held it down with a hoe, chuckling contentedly. Shiver not, my friend, at the callousness of the Ethiopian; that there thing wasn't an infant — it was a ham!

Mary learnt to live on a monotonous bush diet, travelling in a canoe by night, and resting in the beds of water reeds up to 12 ft high by day. Her food consisted of plantain, a banana-like fruit over 12 in. thick; yams, which are hard marrow-sized potatoes; koko yam, the edible tuber of an arum lily; and the thumb-sized seed pods of ochra which were either cooked green or dried and powdered. She also ate fresh and smoked fish, and meat including snails, snakes, crayfish and maggot-like pupae of rhinoceros beetles.

When she arrived home in 1894, Mary brought with her 'a miscellaneous collection of zoological specimens, sufficently large to show the interest she took in the fauna of the countries visited by her'. Dr Guñther was impressed, as he wrote to Mary's friend Mrs J.R. Green:

> Mary Kingsley had not received any training in, or possessed any special qualifications for, Natural History pursuits. . . . She

possessed an extraordinary gift of observation, and she noted a number of facts in the life of animals, of which only a small proportion are recorded in her published writing. ... We finally decided that she should devote her attention to the fishes of the rivers and lakes visited by her. Collecting fishes in spirit is always a laborious and somewhat expensive task, requiring much care and patience. Yet considering the slenderness of her means and her outfit, she succeeded in bringing home a good collection of admirably prepared specimens, a fair proportion of which being new to science, and all valuable additions to any ichthyological museum. Her success was entirely due to her judicious selection of the specimens, and to her indefatigable energy which overcame all obstacles.

She had returned to England only to replenish her supplies of cloth and beads for trading, to earn some money by writing and lecturing — which she had no hesitation in doing — and to seek advice on her specimens from specialists like Dr Günther. She had formed a strong impression of the native African way of life and already was not without criticism of the treatment the Africans received at the hands of the various colonial governments. She had deliberately tried to understand the people and their customs rather than dismissively condemn them as was the fashion of her time. She remembered with glee Livingstone's comment silencing Victorian ladies who ridiculed natives with hoops through their hair. 'Poor things,' he had muttered glancing at the voluminous crinolines, 'they have not learnt yet to put their hoops in the right place.'

The natives in Mary's time were regarded by the missionaries as barely human and certainly not 'elect of God for salvation'. Many missionaries saw their work in Africa as a noble attack on the predominantly Protestant support of slavery. The pseudo-Darwinism taking root at the end of the nineteenth century assumed that the European was superior to the child-like African native. The missionaries were actively hostile to native customs classing them as the works of Satan. Mary determined to show that even the most distasteful behaviour such as cannibalism, human sacrifice, polygamy, twin-murder, slavery, ordeal by poison and the burial of slaves on their master's death, was not the product either of the devil or of an inferior intellect.

On 23rd December 1894, Mary Kingsley left Liverpool on her second and longest journey to West Africa. She embarked on the steamer *Batanga* whose captain was her old friend Captain Murray, and travelled with Lady Macdonald who was going out to join her husband, the Commissioner of the Oil Rivers in Nigeria. They arrived in Calabar to a firework display in welcome of Lady Macdonald.

With the Madonalds, Mary made a brief visit to Fernando Po. Her previous knowledge of the island had been more or less limited to reports from a friend who had lived there. He had returned from Fernando Po with a fever which caused him to shake 'so violently that he dislodged a chandelier and destroyed a valuable tea service and silver tea pot.' She collected marine and freshwater fishes and accurately observed how different the native customs were from those on the mainland. The island's wildlife included little gazelles, small monkeys, porcupines, squirrels, a beautiful golden coloured otter, and gigantic pythons and crocodiles. The latter caused particular concern to Mary on one of her collecting trips by the coast when she had to wade waist deep through several river estuaries. The native's domestic pigs, she noted, 'with no iron living in their noses got adrift and escaped into the interior' where they were destroying crops and the natural habitat. The native Bubis fished (inexpertly in Mary's view) using basket traps and these enabled her to collect specimens. One of their chief sources of food were sea turtle eggs.

On returning to the mainland, Mary stayed with Sir Claude and Lady Macdonald for nearly five months 'puddling about the river and the forest round Duke Town and Creek

Town'. She found a Dr Whitindale who could help her with her collection of fish, and Mr Cooper 'then in charge of the botanical section' in Calabar assisted with the numerous and confusing insects she had collected.

She rejoined the homeward bound *Batanga* to travel north as far as Lagos. Here she had to board the steamer *Benguella* which was to take her south to Gabon. Changing ships at the infamous Lagos Bar, Mary says, 'throws changing [trains] at Clapham Junction into the shade'. They were met at the bar by a branch-boat on which they were to wait for the coastal steamer. Mary was 'most carefully lowered over the side in a chair by the winch' into the dinghy but had to climb a rope-ladder to reach the deck of the branch-boat. They rode out the choppy waters around the sandbar waiting for the *Benguella* which was two days late, an episode Mary did not readily forget.

Mary wanted to collect fishes from a river north of the Congo and her main reason for going to Gabon was to collect in the Ogowé River which had been explored very little before. She had already discovered the fascination and the dangers of travelling through estuarine mangrove swamps in a canoe. She loved the sprawling beauty of the mangroves, changing with every hour of the tide, and:

> the grunts from I know not what, splashes from jumping fish, the peculiar whirr of rushing crabs, and quaint creaking and groaning sounds from the trees and — above all in eeriness — the strange whine and sighing cough of crocodiles.

On two separate occasions a crocodile had chosen, as she puts it,

> to get his front paws over the stern of my canoe, and endeavoured to improve our acquaintance. I had to retire to the bows, to keep the balance right and fetch him a clip on the snout with a paddle.

But, with her usual honesty, Mary adds 'It is no use saying ... I was frightened, for this miserably understates the case.'

Arriving at Libreville in May 1895, Mary was delighted to find that her collecting cases and spirit were passed through customs free because they were for scientific use. '*Vive la France!*' she comments. Her friend Mr Hudson, the agent of a local British company whom she had met at Cabinda (now part of Angola) on her previous visit to Africa, had recommended the Ogowé River for her fish hunting. First she did some collecting in the forest around the town:

> On first entering the grim twilight regions of the forest, you hardly see anything but the vast column-like grey tree stems in their countless thousands around you, and the sparsely vegetated ground beneath. But day by day, as you get trained to your surroundings, you see more and more, and the whole world grows up gradually out of the gloom before your eyes. Snakes, beetles, bats and beasts, people the region that at first seemed lifeless.

It was in these forests that Mary had her first opportuntiy to meet natives belonging to a local tribe called the Fans, in whom she was particularly interested because their habits were reported to include cannibalism.

Mary took the steamer south to the mouth of the Ogowé and then 120 miles upstream to Lambaréné, later the site of Dr Schweitzer's famous hospital. Next she travelled on to Njole, 90 miles further up the river and the last port for steamers. From here Mary set out by canoe with native guides, her portmanteau and her precious trade box at her back, and the French flag 'on an indifferent stick' floating behind her. She trapped fish in a hollow log or a basket, and caught them in nets or on lines or in stockades built across the river. Soon she realised that the fishes she was catching were quite different from those she had seen at Lambaréné.

Canoe travel could be peaceful but collecting specimens was not without its dramas, as she records:

> I was reading: the negroes, always quiet

enough when fishing, were silently carrying on that great African native industry — scratching themselves — so, with our lines over the side, until the middle man hooked a cat-fish. It came on board with an awful grunt, right in the middle of us.

Naturally fearing the fish would escape and a valuable specimen would be lost, Mary's immediate reaction was to shout 'lef em, lef em; hev em for water one time you sons of unsanctified house lizards' and to offer 'such-like valuable service and admonition'.

Mary had set her heart on visiting the rapids on the Ogowé that lay above Njole on a part of the river bordering the Fans' country. The second night they missed the Fan village in which they were to stay and were caught on the rushing river as darkness fell. As Mary recorded in her diary:

> For good all-round inconvenience, give me going full tilt in the dark into the branches of a fallen tree at the pace we were going then — and crash, swish, crackle and there you are, hair being torn out and your clothes ribboned by others, while the wicked river is trying to drag away the canoe from under you.

On reaching a Fan village, they found a collection of palm mat huts with the occupants, almost naked and painted vermilion, dancing enthusiastically. For the next few days they continued to force their small canoe up the rocky path of the Ogowé through towering ravines until they reached a set of deafening rapids which blocked the river completely. One reward of this part of her journey was that at Kondo-Kondo Island, in the middle of the rapids, Mary collected a species of fish which, she later found out, was previously unknown. The trip back down the river to Njole was, if anything, more dangerous and exciting.

She returned to Lambaréné on the steamer. En route the captain insisted on picking up a hippopotamus floating in the river which he had shot on the journey upstream. Amid great excitement, the bloated body was brought on board, and the captain and engineer, coatless and armed with huge butcher knives, prepared to set to work on it. Within two minutes, Mary reported, they came up the ladder 'as if they had been blown up by the boilers bursting, and go as one man for the brandy bottle . . . for remember that hippo had been dead in the warm river-water for more than a week'. She returned to her friends, the Jacots, on a plantation outside Lambaréné, where she gathered information on the local natives and also taught herself to manage a canoe. 'I can honestly and truly say', she concluded, 'that there are only two things I am proud of — one is that Doctor Günther has approved of my fishes, and the other is that I can paddle an Ogowé canoe.'

In late July, Mary set out again by canoe, this time on a broad quiet tributary of the Ogowé. She had time to notice the birds, great hornbills, vivid kingfishers, huge vultures, and egrets. They passed through low lying country with the forest pressing right up to the river bank. As Mary described it:

> Some enormously high columns of green are formed by a sort of climbing plant having taken possession of lightning-struck trees, and in one place it really looks exactly as if someone had spread a great green coverlet over the forest, so as to keep it dry.

On reaching Lake Ncovi, Mary and her companions set out on foot with Fan guides for the Rembwe river. All that saved the weaker members of the party from utter exhaustion was the Fans' appetite. They would sit down every two hours for 'a snack of a pound or so of meat and aguma apiece, followed by a pipe of tobacco'. On these occasions, Mary would walk ahead by herself looking for specimens and animals. In the bottom of a ravine she watched with delight as a group of five elephants rolled and sprayed themselves in a muddy swamp. She was shown a family group of gorillas which she marvelled at but adds as a footnote in her diary that she considered the gorilla the most horrible wild

10.2 Gorillas were portrayed as fierce animals by most nineteenth-century writers (Courtesy of the Bodleian Library, Oxford)

animal she had ever seen. Later they stayed in a Fan village where Mary noticed a vile smell in her hut which she tracked to some bags. Always practical, she shook the contents of one bag out into her hat, 'for fear of losing anything of value', and there were a human hand, three big toes, four eyes, two ears, and other portions of the human frame, mementos of the Fans' recent meal.

The party pushed on through the thick forest swamp and finally reached the Rembwe river. Here Mary bade a 'touching farewell' to her Fan guides. She stayed a few days with Mr Glass, head of Hatton and Cookson's trading post at Agonjo on the river. This so-called factory was the collecting point for local produce such as ebony, ivory and rubber. She learnt more about the Fan customs, including their methods for catching game. The gazelles of the forest, 'with dark grey skins on the upper part of the body, white underneath, and satin-like in sleekness all over . . . the little legs being no thicker than a man's finger, the neck long and the head ornamented with pointed horns and broad round ears', were caught with nets. A party of men and women with trained dogs carrying bells around their necks drove the animals before them through the bush and into the nets.

The Fans were little more ingenious in catching the elephants from which they took the ivory. A herd of elephants would be stalked and edged into a ravine. The area was then fenced in with logs smeared with 'certain things, the smell whereof the elephants detest so much that when they wander up to it, they turn back in disgust'. Weakened by poison in their water supply or by the provision of poison plantain, the animals were finally herded past the hunters sitting in trees who try to shoot them in the back of the neck.

After a delay in finding a suitable canoe for her onward journey, Mary set out on the Rembwe with an African named Obanjo (who preferred to be called Captain Johnson) in a large trading canoe. A few nights later, Mary convinced Obanjo of her helmsmanship and was left to steer the canoe while the crew slept off the fine times they had had ashore. She was always aware of the beauty of her surroundings, and this moonlit journey particularly impressed her, as this passage shows:

> The great, black, winding river with a pathway in its midst of frosted silver where the moonlight struck it: on each side the ink-black mangrove walls, and above them the bank of star and moonlit heavens that the wall of mangroves allowed one to see. . . . On the second night, towards dawn, I had the great joy of seeing Mount Okoneto, away to the S.W., first showing moonlit, and then taking the colours of the dawn before they reached us down below. Ah me! give me a West African river and a canoe for sheer good pleasure.

When they reached the point at which the Rembwe, with several other local rivers, emptied itself into the Gabon estuary, canoeing became more tricky. At one stage, Mary's portmanteau and other belongings went temporarily overboard. However, they arrived safely in Glass on the north side of the estuary. Mary went immediately to find her friend Dr Nassau 'to discourse . . . on Fetish'. Dr Nassau had worked with the American Presbyterian Mission on Corisco Island, 20 miles offshore, which had given him the doubtful privilege of being 'the first white man to send home gorillas' brains'. He told Mary of lakes in the centre of the island which contained quantities of fish; Corisco became her next destination. Dr Nassau obtained a sailing vessel for her in Libreville and, propped comfortably on her collecting box, Mary set off for Corisco.

The women of Corisco, it seemed, were the only ones able to catch the lake fish. Mary waited impatiently for a few days while the women were preparing their fishing baskets. These baskets were shaped like pillows with one side open. The fish were driven towards a line of women by the shore and trapped with much shouting and splashing. To Mary's immense disappointment, the only fish in the lake turned out to be the familiar mud-fish, a species she had collected almost everywhere else.

The return journey to the mainland was made difficult by rough seas, mist and finally by a pair of whales breaching within 30ft. of the boat and nearly swamping it. Once safely back in Libreville, Mary had time to reflect on all she had learnt since setting out in May. The Ogowé River had been scarcely explored until 30 years before her visit when French traders began to trace its course inland. The first survey of the upper reaches of the river was made by the explorer de Brazza, and the area was also visited by Paul du Chaillu, well known to Mary's contemporaries in England for his exciting, but actually ridiculous, reports of gorillas. Mary had collected a huge amount of information on native language, legends, customs and beliefs during her journey. In her diary she had kept detailed records of charms, sacrificial rites, witchcraft, reincarnation beliefs, twin murder, burial customs, mourning, inheritance laws, punishment of offenders, witches, albino deities, surf spirits and wandering souls. The subjects occupy over a hundred pages in her book *Travels in West Africa*.

In the middle of this discourse, Mary reports a meeting she had with a leopard.

> Thus I once came upon a leopard. I had got caught in a tornado in a dense forest. The massive, mighty trees were waving like a wheat-field in autumn . . . the tornado shrieked like ten thousand vengeful demons . . . the fierce rain came in a roar, tearing to shreds the leaves and blossoms and deluging everything. I was making bad weather of it, and climbing up over a lot of rocks out of a gully bottom where I had been half drowned in a stream, and on getting my head to the level of a block of rock I observed right in front of my eyes, broadside on, maybe a yard off, certainly not more, a big leopard. He was crouching

10.3 Mary had several close encounters with leopards

on the ground, with his magnificent head thrown back and his eyes shut. His forepaws were spread out in front of him and he lashed the ground with his tail, and I grieve to say, in face of that awful danger — I don't mean me, but the tornado — that depraved creature swore, softly, but repeatedly and profoundly.

Mary hid under the rocks and remained there for what felt like ages, but was actually about 20 minutes, listening to 'his observation on the weather, and the flip-flap of his tail on the ground', until she looked out and he was gone.

Mary seems to have had a fascination for leopards, or vice versa. On another occasion she was awoken in the night by a dog being attacked by a leopard. She threw two native stools into the whirling mass of animals and broke up the fight. For a brief moment the leopard crouched ready to spring staring at her with 'its great, beautiful, lambent eyes', until with considerable common sense she grabbed an earthenware water cooler and flung it at the animal's head. 'Twenty minutes after', she adds, 'people began to drop in cautiously and inquire if anything was the matter, and I civilly asked them to go and ask the leopard in the bush, but they firmly refused.' The dog had had its shoulder split open as if from a cutlass blow.

After a few weeks more in Gabon, Mary left for Cameroon. Ever since she had first visited this coast in 1893, she had an ambition to climb the great Peak of Cameroon, Mungo Mah Lobeh or the Throne of Thunder, a mountain over 13,000ft. high. It is hard to see where this idea came from as she herself admitted that 'there's next to no fish on [mountains] in West Africa, and precious little good rank fetish, as the population on them is rather sparse.'

The first man ever to reach the summit of the Peak was Sir Richard Burton (see chapter 6), but he had climbed up from the western side. The eastern ascent, which Mary now proposed to attempt, had been made only once, a few weeks before her arrival. Why she chose this difficult route remains a mystery although, as she later claimed in a lecture to the Royal Scottish Geological Society, she was anxious to observe the topography of the mountain, in other words with characteristic curiosity she simply wanted to see what it was like at the top!

She set out on 20th September from Victoria on the coast, and six days later after incredible hardships she stood on the summit, inevitably shrouded in blinding, stinging rain. Keeping her small party of guides and natives from turning back on the way up had taxed even Mary's ingenuity. It had rained almost continuously, changing the ground beneath them to deep mud or streaming floods. They had pushed forwards and upwards through a flashing tornado and the bewildering mountain mists, the tropical rain forest seeming like a vast hall, its columns covered in dark green moss and delicate ferns. The return journey was made at almost literally breakneck speed so eager were the men to get off the mountain. Mary stayed with her friend Herr von Lucke in Victoria while waiting for the steamer *Nachtigal* which would take her to Calabar. There, in October 1895, she boarded a ship bound for Sierra Leone and home.

On her return, she settled into her London house at 100 Addison Road, Kensington, and began to write letters, lectures, articles and a book. A photograph taken about this time shows Mary with a half smile on her face sitting in a rocking chair. She wears a pillbox hat low on her brow and perched on a bun, and a black dress with puff sleeves tight at the arm and wrist.

By the time the account of her adventures *Travels in West Africa* appeared in 1897, Mary was already something of a celebrity. She had lost no time in airing her views in lectures and articles on how the native Africans in British territories should be treated. She publicly opposed a controversial tax on native huts proposed by the government. This brought her unwanted notoriety and made her unpopular with African authorities. As her friend Stephen Gwynn wrote of Mary: 'Where she saw injustice done to native Africans, and done by her own country, in the name of civilisation, then she could not sit still.' Without knowledge of the people and their customs, Mary was convinced, there would be no justice, and without justice, the natives would always be treated dishonourably.

In 1898, Mary started writing her second book *West African Studies* which was published the following year. This ran to 633 pages in its original edition and had 193 appendices. But, to quote Stephen Gwynn again:

> Whatever else is to be remembered about Mary Kingsley, this passage ought never to be forgotten: All we need look to is justice. Love for our fellow-men, pity, charity, mercy, we need not bother our heads about, so long as we are just.

The book was really a soapbox for Mary's appeal for fair government in the British colonies and protectorates in Africa. It is for this campaign for the welfare of the natives that Mary Kingsley is remembered best. A review of *West African Studies* in the *Manchester Guardian* newspaper stated it thus: 'Miss Kingsley has followed no school of West African policy, but, if we mistake not, she has laid the foundation of one.'

Both Mary's books are lengthy in style, and often confused by side-references to miscellaneous facts she picked up along her way. But they remain, even almost a century later, a most readable account of her travels and of native West African customs; and an impassioned and well-informed plea for understanding of the African. In her biographical anthology of the Kingsleys, Elspeth Huxley says of Mary: 'Judged by modern rather than Victorian standards she was, as a writer, perhaps the greatest Kingsley of them all.'

Mary's other important contribution, of course, was her collection of specimens. These were passed to the British Museum on her return from Africa. Early in 1896 she wrote in a letter to a friend:

> You should just hear Dr Günther at present because I have not been over there helping him with decayed fish. I *did* go over only the other day and worked for hours, and stank so that people noticed a 'strange smell' at luncheon and dinner parties I was present at for a week after. It is a cruel world.

In February, she received the verdict on her collections from Dr Günther. She had brought home 18 species of reptile including a lizard collected at Lambaréné which was only the second specimen ever known — one the British Museum 'have been waiting for for ten years', said Mary. Among her 65 species of fish were eight new species, three of which were named *kingsleyae* after her, and many other species which provided valuable new information on the distribution of African fishes. Mary herself concluded: 'I can now say Dr Günther is satisfied with my work: he wants more at once of course, and I am glad, for I was beginning to fear I was an utter windbag.'

Dr Günther added a paper of his own on the reptiles and fishes Mary had collected, published in the *Annals and Magazine of Natural History* for 1896, as an appendix to her *Travels in West Africa*. Her insect specimens collected at Lambaréné on the Ogowé River were examined by W. H. F. Kirby at the British Museum, who added his own appendix to the book. He found eight new species in the collection, including two in entirely new genera. Also listed in the appendix are 26 molluscs whose shells Mary had collected mainly on Corisco Island (while waiting for those women to get their fishing baskets ready), and 36 species of plants found on the Ogowé and on the Cameroon mountains.

Mary longed to return to Africa and it was a need that did not abate with time, despite the feeling that her duty lay in enlightening English society. When the Boer War broke out, Mary immediately made it her excuse to visit South Africa, alleging she would also collect fish for Dr Günther in the Orange River. To her delight, while passing through Cape Town in March 1900, she met the author Rudyard Kipling and his wife. She obtained work in a fever camp at Simonstown, typically choosing to nurse Boer prisoners rather than British soldiers. Here she worked for two months with her usual fortitude, and compassion for the sick.

Conditions were far from ideal and in this the first British war since the Crimea, little had been learnt of Florence Nightingale's methods. Even in this short time, according to a doctor at Simonstown, Mary's 'quiet dignity, her clear, capable mind, her practical ability, her unfailing good humour, her tenderness and sympathy, won for her here, as elsewhere, not only respect but affection from patients and colleagues'.

On 3rd June 1900, Mary Kingsley died of a heart attack caused by a severe bout of enteric fever. Her funeral was held at sea in accordance with her wishes, and was one of the most impressive events Simonstown had ever witnessed. The world had lost 'a personality admirable for its integrity, memorable and lovable for its oddity and variety — its tenderness and ferocity, its laughter and its pity.'

Mary Kingsley's work did not die with her. Within a few months of her death, what is now the Royal African Society was founded. Article I of its Constitution, in commemoration of Mary's work, promised to investigate all African races and to secure their future.

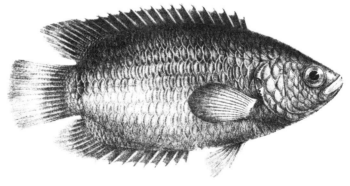

10.4 Ctenopoma kingsleyae, *one of the three new species of fish Mary collected in West Africa*

CHAPTER ELEVEN
Epilogue: The End of an Era

11.1 Charles Darwin in 1847, eleven years after he returned from his journey around the world on the Beagle *(Courtesy of the BBC Hulton Picture Library)*

By the last decades of the nineteenth century, the crazes for natural history had been abandoned in Britain. The Romantic period in both art and literature was being replaced by a new realism; and the countryside was no longer regarded as the place for recreational natural history collecting. Meanwhile, the scientific community was embroiled in a fierce controversy over Charles Darwin's ideas on evolution.

Although the existence of evolution had been proposed earlier in the nineteenth century, Darwin's suggestion that the seemingly random process of natural selection was the main way in which species evolved was met with dismay. Furthermore, many found especially unacceptable the idea that man was simply part of this process and that he had not been created in the Garden of Eden. Darwin's critics misinterpreted his conclusion that apes and man shared a common ancestor at one stage in their respective evolutionary histories, and claimed that man was descended from the apes. Not surprisingly, these new ideas had a profound effect upon the general perception of man's place in the natural world.

Darwin's book *The origin of species by means of natural selection* was published in November 1859. The 1250 copies printed sold out in one day and the book did not appear again until two months later when 2000 copies of a second edition had been printed. The initial reactions to Darwin's theories were cautious and uncritical. By chance Darwin's great friend and advocate Thomas Huxley was asked to write a review of the book for *The Times* newspaper, and he expressed a confident and enthusiastic opinion. It was not until the second edition of *The origin* appeared that church leaders began to see sacrilege and sedition in Darwin's arguments.

Matters came to a head at the now famous British Association meeting in June 1860 which was held in Oxford. Darwin was too ill to attend but Huxley confronted first the notable paleontologist Sir Richard Owen and then the Bishop of Oxford, 'Soapy Sam' Wilberforce. Having crossed swords in public with both the traditional scientific community and the Church, Darwinism, as it came to be known, was ensured publicity and continuing debate.

Although incorrect, the idea of apes as ancestors was what many people found hardest to accept. The arrival of Paul du Chaillu from Africa in 1861 with no less than 20 stuffed gorillas, their skins and skeletons did nothing to reassure the public who flocked to his lectures where these specimens, the first gorillas ever seen in Europe, were dramatically displayed. The apparent separation of apes and man was diminished even further when Huxley showed that the single feature of the human brain (the possession of a *hippocampus minor*) which, it had been suggested by Owen and others, was unique to man, was present also in the anthropoid apes. The controversy was still raging when Darwin published *The descent of man* in 1871, but it was not until the rediscovery of genetics in 1900 that the exact mechanism of evolution was elucidated.

Darwinism and the controversy which surrounded it had a marked effect upon amateur naturalists and collectors. Darwin had shown that species were not the immutable, precisely defined units which they had been considered previously. The collector's goal of obtaining a complete set of all species was made less meaningful by the new concept of the continuous evolution of the natural world. Instead of chasing more species for their collections, many naturalists turned to studying live animals in their natural surroundings. This gave birth to the discipline of ecology, or oecology as it was known initially, to the study of animal behaviour and, eventually, to the wildlife conservation movement. These were aided by the increasing availability of prismatic binoculars, cameras, flashlights and even telephoto lenses during the 1880s and 1890s. The brothers Cherry and Richard Kearton and Ludwig Koch were among the pioneers of wildlife photography and sound recording respectively at about this time.

All of these changes undermined the role of the travelling naturalists. Their specimens were no longer eagerly sought after for museum collections, and the prestige and incentive to making private collections was dying. The debate over Darwinism was conducted largely by the academic and religious communities, and only the more generalised, or sensational, aspects of it filtered down to the public. With little or no academic training, and a fund of detailed experience and knowledge in which no one seemed interested, many of the naturalists must have felt very out of touch with society on their return to Britain.

However, travelling naturalists continued to exist even after the turn of the century, although they developed along different lines from their predecessors. They were academically educated, and often trained in the various specialised fields of biology. The study of ecology grew, and the new biologists began to travel abroad to examine exotic ecosystems using the theories and techniques they had worked out at home. Today active field scientists are to be found in every continent from the ice caps of Antarctica to the tropical forests of Africa.

As photography expanded into cinematography, wildlife filming became an important tool in the hands of both educators and research workers. When subjects nearer home were exhausted, travel became inevitable until an expedition overseas formed a normal part of the work of many wildlife film makers, as modern films and television programmes about wildlife show. These twentieth-century 'travelling naturalists' may seem a long way removed from the wandering journeys and arboreal pursuits of Charles Waterton, but he could rightly be considered their ancestor.

As we approach the end of this century in which the unexplored regions of the earth have dwindled to almost nothing, it is easy to regret that the days of such adventures are passed. The scope for exploration and the discovery of new species may have diminished, but the challenges facing us remain intimidating. Many more generations of naturalists will be needed both at home and abroad if we are to understand and conserve what we have for our own time and for times to come.

Select Bibliography

Aldington, R. *The strange life of Charles Waterton* (Evans, London, 1949)

Anon. 'Obituary of Admiral Sir F.L. McClintock', *The Irish Naturalist*, vol. 17 (1908) pp. 42-3.

Anon. 'Obituary. Mr. Howard Saunders', *Ibis*, 9th Series, vol. 2 (1908) pp. 169–72.

Anon. 'Memoir of the late Henry Seebohm, F.L.S., F.Z.S., Secretary to the Royal Geographical Society', *The Zoologist*, vol. 20 (1896) pp. 10–14

Anon. 'Obituary — Mr. H. Seebohm', *Ibis*, vol. 2 (1896) pp. 159–62

Anon. 'Obituary — Mr. Henry Seebohm', *Auk*, vol. 13 (1896) pp. 96–7

Barber, L. *The heyday of natural history. 1820-70* (Cape, London, 1980)

Bates, H.W. *The naturalist on the river Amazonas* (Murray, London, 1863)

Bird, I. L. *A lady's life in the Rocky Mountains* (Murray, London, 1879)

Burton, R. F. *First footsteps in East Africa* (Longmans, London, 1856)

——— *The lake regions of central Africa: a picture of exploration* (London, 1860)

——— & MacQueen, J. *The Nile basin* (Tinsley Brothers, London, 1864)

Campbell, O. W. *Mary Kingsley: a Victorian in the jungle* (Methuen, London, 1957)

Chapman, A. 'A memoir of Howard Saunders', *British Birds*, vol. 1 (1907) pp. 197–201

Dresser, H. E. *History of the birds of Europe* (London, 1871–81)

Du Chaillu, P. *The world of the great forest* (Murray, London, 1901)

Elliston Allen, D. *The naturalist in Britain, a social history* (Penguin, London, 1978)

Glynn, R. *Mary Kingsley in Africa. Her travels in Nigeria and Equatorial Africa told largely in her own words* (Harrap, London, 1956)

Gosse, P. G. H. *The Squire of Walton Hall* (Cassells, London, 1940)

Gosse, P. H. *The aquarium: an unveiling of the wonders of the deep sea* (Van Voorst, London, 1856)

——— *Actinologia Britannica. A history of the British sea-anemones and corals* (Van Voorst, London, 1860)

Grant, J. A. *A walk across Africa, or domestic scenes from my Nile journal* (Blackwood & Sons, London, 1864)

Green, W. S. 'Report on a journey among the New Zealand glaciers in 1882', *Proceedings of the Royal Irish Academy*, Series 2, vol. 3 — Science (1883) pp. 642–54

——— *The high alps of New Zealand* (Macmillan, London, 1883)

——— *Among the Selkirk glaciers* (Macmillan, London, 1890)

——— 'Notes on Rockall island and bank, with an account of the petrology of Rockall, and of its winds, currents, etc.: with reports on the ornithology, the invertebrate fauna of the bank, and on its previous history' *Royal Irish Academy Transactions*, vol. 31, part 3 (1896) pp. 39–47

Günther, A. E. *A century of zoology at the British Museum through the lives of two keepers, 1815–1914* (Dawsons, London, 1975)

Gwynn, S. *The life of Mary Kingsley* (Macmillan, London, 1933)

Haughton, S. 'On the fossils brought home from arctic regions in 1859, by Captain Sir F. L. McClintock', *Royal Dublin Society Journal*, vol. 3 (1860) pp. 53–8

——— 'Geological account of the arctic archipelago, drawn up from the specimens collected by Captain F. L. McClintock, R.N., from 1849 to 1859', *Geological Society of Dublin Journal*, vol. 8 (1860) pp. 196–221

——— 'The voyage of the "Fox" in the arctic seas', *Dublin University Magazine*, vol. 55 (1860) pp. 208–21

Huxley, E. J. *The Kingsleys: A biographical anthology* (Allen & Unwin, London, 1973)

Jenkins, A. C. *The naturalists: Pioneers of natural history* (Hamilton, London, 1978)

Select Bibliography

Kastner, J. *A world of naturalists* (Murray, London, 1978)

Kingsley, M. K. *Travels in West Africa* (Macmillan, London, 1897)

―――― *West African studies* (Macmillan, London, 1899)

Kirwan, L. P. *A history of polar exploration* (Penguin, London, 1962)

Maitland, A. *Speke* (Constable, London, 1971)

McClintock, F. L. 'Reminiscences of arctic ice-travel in search of Sir John Franklin and his companions', *Royal Dublin Society Journal*, vol. 1 (1857) pp. 183–250

―――― 'Report of scientific researches made during the late arctic voyage of the yacht "Fox", in search of the Franklin Expedition', *Royal Society of London Proceedings*, vol. 10 (1859–60) pp. 148–51

―――― *A narrative of the discovery of the fate of Sir John Franklin and his companions. The voyage of the 'Fox' in arctic seas* (London, 1860)

Markham, C. *Life of Admiral Sir Leopold McClintock* (Murray, London, 1909)

Moorhead, A. *The White Nile* (Hamilton, London, 1960)

Ogilvie-Grant, W. R. 'Obituary. Howard Saunders', *The Zoologist*, vol. 2 (1907) pp. 436–8

Praeger, R. L. *The way that I went. An Irishman in Ireland* (Hodges, Figgis, Dublin, 1937)

―――― *Some Irish naturalists. A biographical notebook* (Dundalgon Press, Dundalk, 1949)

Saunders, H. 'Ornithological rambles in Spain', *Ibis*, vol. 5 (1869) pp. 170–86

―――― 'Across the Andes and down the Amazonas', *The Field, the Country Gentleman's Newspaper*, 19th February 1881, pp. 244 et seq.

―――― *An illustrated manual of British birds* (London, 1889)

Scharf, R. F. 'Obituary of William Spottswood Green', *Irish Naturalist*, vol. 28 (1919) pp. 81–4

Sclater, P. L. 'On the mammals collected and observed by Captain J. H. Speke during the East-African Expedition', *Proceedings of the Zoological Society of London* (1864) pp. 98–106

Seebohm, H. *The birds of Siberia. A record of a naturalist's visits to the valleys of the Petchora and Yenesei* (Murray, London, 1901)

―――― *Coloured figures of eggs of British birds with descriptive notices* (Sheffield, 1896)

―――― & Sharpe, R. B. *A monograph of the Turdidae or family of thrushes* (London, 1891–1902)

Sitwell, E. 'Charles Waterton: the South American wanderers' in *The English Eccentrics* (London, 1933)

Speke, J. H. *Journal of the discovery of the source of the Nile* (Blackwood & Sons, London, 1863)

―――― *What led to the discovery of the source of the Nile* (Blackwood & Sons, London, 1864)

Walker, D. 'Notes on the zoology of the last arctic expedition under Captain Sir F. L. McClintock, R.N.' *Royal Dublin Society Journal*, vol. 3 (1860) pp. 61–77

Wallace, A. R. *A narrative of travels on the Amazon and Rio Negro* (Ward & Lock, London, 1853)

―――― *My life* (Chapman & Hall, London, 1905)

Waterton, C. *Wanderings in South America, the north-west of the United States of America and the Antilles in the years 1812, 1816, 1820, 1824* (London, 1825)

―――― *Essays on Natural History* (Warne, London, 1871)

Watson, G. *Charles Waterton. 1782–1865. Traveller and naturalist* (Wakefield Art Gallery and Museums, 1982)

Went, A. E. J. 'William Spottswood Green', *Royal Dublin Society, Scientific Proceedings*, Series B, vol. 2, no. 3 (1967) pp. 17–35

Wood, Rev. J. G. *The illustrated natural history* (Routledge, London, 1861–3)

Postscript

There is an extraordinary sequel to the story of Franklin's expedition and its disappearance (Chapter 4). One hundred and thirty four years after McClintock and McClure discovered the remains of Franklin's winter base on Beechey Island, a modern expedition visited and investigated the site. As this book was going to press, Dr Owen Beattie of the University of Alberta, led a team of anthropologists to Beechey Island in the hope of discovering more about the fate of the Franklin expedition.

Nineteenth-century visitors to Beechey Island had found three roughly carved tombstones marking the graves of Leading Stoker (or Petty Officer) John Torrington of the *Terror*, and of a seaman and a marine from the *Erebus*. Beattie's team disinterred Torrington's corpse and found it perfectly preserved by its tomb of permafrost. Photographs show the man's youthful and undistorted face, his effeminate hands, and almost uncreased pinstriped navy shirt and trousers. Beattie's hopes that analysis of tissue and organ samples taken from Torrington's body before it was re-buried will indicate an exact cause of death.

Similar analyses of bones found by Beattie's team on an earlier expedition to King William Island suggest that Franklin's men may have suffered from lead poisoning, possibly as a result of eating meat from cans sealed with lead, and from scurvy. Another speculation, which immediately caught the ear of the world's press, was that curious slash marks on the bones indicated cannibalism by other members of the expedition. Equally probable, however, is the explanation that these were made by scavenging Arctic foxes.

Index

Note: Colour plates are denoted by C.

Admiralty, the 23, 41, 48
albatross 12-14
alligator 23, 68, 70, 101
Amazon kingfisher 101
Amazon River 99-103; *see also* Bates, Henry
American robin C15
anteater 31, 68
ants: bull-dog 123; driver 138; sauba 63
aquaria, private and public 15
Archangel, early history 112
Arctic hare C2; *see also* McClintock, William
Asiatic Society 80
Asulkan Pass 131
Audubon, John James 16, 18, 38, C1, C20
auks: ancient murrelet C1; crested auklet C1; guillemots 38, 46, 50-1; least auklet C1; little auk 46, 50-1; razorbill 38; rhinoceros auklet C1
Austin, Captain 46
Australia 16, 122-3

Baffin, William 41
Banks, Joseph 24, 28, 33-4
barn owl 31
Barrington, R.M. 132
Barrow, John 41
Bates, Henry 17, 23-5, 94, 101, C6, C13; life and work 60-72, appearance 66, correspondence with Darwin 72, death 72, early life 61-2, expedition to Brazil 62-71, map of journeys 60
Batesian mimicry 72
bats 31, 63
bears: grizzly 131, C20; polar 49, 51

Beechey Island 42, 46, 51; *see also Postscript*
bee-eater C3
Belcher, Edward 47-8
Belgian Congo 138
bell-bird 31, 33, 123
binoculars 148
binomial nomenclature 12; views of Seebohm 118; views of Waterton 39
Bird, Isabella 21, 135, C21
bird-eating spider 66-7
birds of paradise 15
Blakiston, Thomas 36
blanc bok 87
Blyth, Edward 80, 82
boat-billed heron 29
Bodleian Library, Oxford C7-9, C13, C17
books: illustrations 19, C7-8; methods of illustration 14, C9; printing of 14
Boss, Emil 122-7 *passim*
Botanical Society of London 13
Brazil 32, 99-103; *see also* Bates, Henry
brimstone butterflies 63
British Association 122, 130, 148
British Association for the Advancement of Science 13
British Guiana *see* Waterton, Charles
British Museum 23-4, 118, 132, 137, 145-6
British Ornithologists' Club 93, 119
British Ornithologists' Union 13, 93
bronze-winged pigeon 123
buffalo 83-4, 90
Burton, Isabel 76
Burton, Richard 11, 16, 19, 25, 75, 144, C10; early life 78; on Speke's appearance and character 79
bush buck 90

bustard 103
butterflies 63, 69; family Ithomidae 72; genus Heliconius 17; genus Morpho 69; *see also* mimicry

caiman 31, 34-6, 39
Calabar 138-9, 145
Cameroun 138, 144-5
Canada 16, 18, 36; *see also* McClintock, William
Cape Bird 52, 56
capercaillie 109
Cape York 46, 50
capybara 99, 101
caribou 130-1; *see also* reindeer
chestnut-collared swifts 97
China 16, 18, 110
classification, system of *see* binomial nomenclature
closet naturalist 15, 93
clothing 20-1
collecting: at seaside 14; crazes for 14; insects 22; preservation of specimens 22
Cook, Captain 28
corrosive sublimate (mercuric chloride) 34
cotinga 33, 65
coypu 99
Crimean War 43, 48, 82
curare 30
curassow 100

Darwin, Charles 15, 22-4, 62, 101, 123, 147-8
dolphin 65, 101
Dresser, Henry Eeles 110, C3-4, C14
Du Chaillu, Paul 18, 143, 148
ducks: blue 125; eider 49-50; paradise 125; teal C4
Dundalgon Press, Dundalk 128
Dürer, Albrecht 14
Dutch Guiana 33

153

eagles: Bonelli's 103-4, C14; harpy 101; white-tailed 110, 131
East Africa 16; *see also* Speke, John Hanning
ecology 148
Edinburgh Society 13
Edmonstone, Anne Mary 29, 38
Edmonstone, Charles 29
Edward Grey Institute, Oxford 107, C1, C3-6, C14-16, C18
eighteenth century, study of natural history 12
El Dorado 30, 32
Entomological Society 12, 72
equipment 20, 22

Fernando Po 139
fetish *see* Kingsley, Mary
fish: Mary Kingsley's collection 138-9, 146; wrasse C9
flamingo 101
florikan sandgrouse 80, 88
food 48, 99-100, 124, 138
fossils *see* McClintock, William
Franklin, Jane 43, 48, 56
Franklin, John 16, 41-2, 52-6; *see also Postscript*
French Gabon *see* Kingsley, Mary
French Guiana 32
frigatebird, magnificent 32
frogs 31, 63

Gabon *see* Kingsley, Mary
gannet 103
glossy ibis 95
goatsuckers 31, 88-9
goose: bean 110; brent 56
Gordon Cummings, Constance 135
gorilla 18, 141-3, 148, C18
Gosse, Philip 14, 18, C7-9
Gould, John 88, C6, C18
Grant, James 13, 16, 90
Grant's gazelle 13, 86, 90
great crested grebe 49
great egrets 95
Green, Charles 132
Green, William Spottswood 16, 18, 20, 24, C17, C20; life and work 120-33, death 133,
dredging expeditions 128-30, 132, early life 121, expedition to New Zealand 122-7, expedition to Rockall 132, expedition to Selkirk Rockies 130-2, first ascent of Mount Cook 126-7, Irish fishing industry 129, map of journeys 120, portrait 128
green jay 97
Greenland 49-51
green mitred parakeets 96
grey plover 110-11
Guillemard, Dr 24, 136-7
guinea-fowl 88
guinea pig 29, 99
gulls: Andean 95; black-billed 125-6; glaucous 51, 56; grey-hooded 69; herring 110; ivory 51, 57-8; kittiwake 46, 132
Günther, Albert 16, 23, 118, 132, 137-8, 145-6
Guyana *see* Waterton, Charles
Gwynn, Stephen 132, 145

Haddon, Professor A.C. 128, 132
hartebeest 88
Harvie-Brown, John 132; *see also* Seebohm, Henry
Haughton, Samuel 47, 57, 128
Herne, George 79-82
hippopotamus 18, 82, 84-5, 88, 141
hoary marmot 131
hoatzin 65, 101-2
Hobson, William 48-56 *passim*
Hooker, Joseph 16, 23, 90, 127
Hooker, William 16
hoopoe 104
hornbill 141
horned screamer 29-30
hummingbirds 33, 37, 66, 97
Huxley, Elspeth 145
Huxley, Thomas 16, 24, 148, C11
hyena 88

Ibis 94, 104-5
Iceland 23
Incas 95, 97
India 16, 77-8
insects 12, 22, 25; Bates's collection 60-72; beetles C13; cicadas 63; mosquitoes 25, 111; rhinoceros beetle 37; *see also* ants, butterflies, moths
Ireland 43-4, 47, 58; *see also* Green, Charles

jaguar 31, 98, 101, C19

Kamrasi, King of Unyoro 89
Kaufman, Ulrich 122-7 *passim*
Kearton, Cherry and Richard 148
Keulemans, J.G. 119
Kew botanical gardens 16, 90, 127
Khartoum 16, 89-90
killdeer 95
Kingsley, Charles 136
Kingsley, George 136, 137
Kingsley, Mary 17, 20, 22-5; life and work 134-46, appearance in 1895 135, 145, death 146, early life 135-6, first visit to Africa 138-9, map of journeys 134, portrait 135, preparations for travel 137, second visit to Africa 139-45, views on West African society 139, 145
kiskadee flycatcher 31
Koch, Ludwig 148
Krapf, Johann 79

Lake Victoria 76, 83
leopard 80, 143-4
Linnean nomenclature 12
Linnean Society 12, 72, 94, 119
Livingstone, David 90, 139
Livingstone's eland 86, 90
lizards 63, 81
Low, Frederick 44, 94
lynx 80

macaws 30; hyacinthine 65; scarlet and blue 68
McClintock, Leopold 16, 128; life and work 40-59, death 59, discovery of Bellot Strait 46, fourth Arctic expedition

48-6, early life 43-4, first Arctic expedition 45-6, *Fox* expedition 48-56, in West Indies 57-9, map of journeys 40; portrait 43, second Arctic expedition 46-7, third Arctic expedition 47-8, *see also Postscript*
McClintock, William 44
McClure, Captain 46-8
Macgillivray, John 16
Macgillivray, William 16
McQueen, James 85, 90
Madeira 16
Maitland, Alexander 85
manatee 100
Markham, Clements 45, 59
marvelous spatuletail 97
microscopes 14, 22
mimicry 72
moa 124
monkeys: howler 31, 37, 98; monk saki 61; scarlet-faced 61; spider 69
Moore, Norman 38-9
Moorhead, Alan 78-9, 88
moths 66, 93, 98
Mtesa, King of Buganda 88
Murchison, Roderick Impey 41-2, 76, 85
Murray, Captain 138-9

National Portrait Gallery, London 75, C10-12
native tribes 11; Aaquaruna 97; Eskimo 49, 53; Fan 140-2; Macoushi 31; Ostiak 115; Samoyede 108-9; Tushaua 71; Yurak 17; *see also* Saunders, Howard
natural selection 147
New Zealand 123-7
New Zealand harrier 124
Nondescript, the 37
North, Marianne 135
Northwest Passage 41-2, 55

Ord, George 38, C12
osprey 131

paca 100
pallah bok 90
Palliser Expedition 36
parrots: kea 125-6, C18; *see also* macaws, toucans
Peale, Charles Wilson 36
peccary 31, 100
pennant-winged nightjar 88-9
Peru 44, 94-102
Petersen, Carl 48, 50
petrels: black-capped 58; cape pigeon 124; great shearwater 132; Jamaica 58; prion 122; *see also* albatross
Phillip's dik-dik 80; *see also* Salt's dik-dik
photography 19, 24, 85, 148, 149
Praeger, Robert Lloyd 129, 130-2
Prethrick, John 16, 89
puma 96

Rae, John 43, 48
red-legged partridge 103
redstart 110
reindeer 47, 49-50, 109-10; *see also* caribou
rhinoceros 80, 86-7, 90
Ripon Falls 76, 88
Ross, James Clarke 41, 45
Royal African Society 146
Royal Dublin Society 48, 56-8, 129, 131
Royal Geographic Society 23, 41-2, 56, 58, 72, 79, 82-3, 88-90, 94, 119, 128
Royal Irish Academy 122, 128-30, 132
Royal Scottish Geological Society 144
Royal Society 57, 72
Rumanika, King of Karagwe 87
Russia *see* Seebohm

sable antelope 87
saltiana antelope 80
Salt's dik-dik 80, 90
Saunders, Howard 17, 20, 24, 44, C3-4, C14, C19; life and work 92-105, early life 94, expedition to South America 94-102, map of journeys 92, visit to Spain in 1867 103-4
scarlet ibis 29, 31

Sclater, P.L. 23, 85, 90, 94
Seebohm, Henry 17, 23-4, C15-16; life and work 106-19, British Museum catalogues 118-19, early life 107, first visit to Siberia 108-12, map of journeys 106, portrait 107, second visit to Siberia 113-18, views on binomial nomenclature 118
Selkirk Mountains 130-2
serval 89
Sharpe, R. Bowdler 113, 119
Siberian jay 109
Sierra Leone 88, 138, 140, 145
silver-beaked tanager 63
sloth 30-1, 37
Smith, Sydney 37-8
Smyth, William 44-5, 94
snakes 11, 30, 98
snow bunting 45, 109, 114
Somerset Island 42, 45, 46, 51
Southern Alps *see* Green
species: naming *see* binomial nomenclature; treatment by closet naturalists 15
specimens: preparation of 32, 33-5, 39, 63, 65, 85-7, 104, 145; treatment of 23
Speke, John Hanning 16, 19, 23-5, 75, C13; life and work 74-90, death 76, early life 77-8, first African expedition 79-82, map of journeys 74, portrait 75, second African expedition 82-5, third African expedition 85-90
Speke's gazelle 80
spiders 66-7
Stanley, Henry 90
starling 103
Stejneger, L. 119
Stonyhurst College 28, 39
Stroyan, William 79, 82
Surinam 33
Swainson, William 38
swallowtail butterflies 63
swans: Bewick's 110; whistling 49
Swanzy, Henry 130-1
Swinhoe, Robert 16, 110, 118

Index

takahe 124
tapir 31, 101
taxidermy 33-5; *see also* specimens
terns: black skimmer 101; large-billed 69; yellow-billed 101
Thomas' kob 88
tinamou 30
topi 86
toucans 31, 33, 64; Cuvier's C6
Transit of Venus expedition 105
travel: equipment 20, 22; funding of 24; preparations for 20; reasons for 20
Trinity College, Dublin 47, 96, 121
tropicbird, red-billed 32-3
troupial 33, 63
turtles 70, 99-100, 149

umbrella bird 70
United States 18, 21, 24, 27, 57
Ural Mountains 107, 113
Ussher, Robert 132

vanessid butterflies 63
Von Haast, Julius 123-4
vultures 122; black 97; Egyptian 103; king 97; lammergeier 103; turkey 95

wading birds *see* Seebohm, Henry
Wakefield Art Galleries and Museum 37, 39, C21
Walker, David 48-56 *passim*
wallaby 121, 123
Wallace, Alfred 18, 23-4, 62, 69
Wardian display case 15
waterbuck: common (sitatunga) 86-7, 90; defassa 89-90; ellipsiprymnus 88
Waterton, Charles 11, 17, 24, C5, C12, C21; life and work 26-39, appearance at 82, 38, capture of caiman 34-6, death 39, early life 27-8, first journey 28-32, fourth journey 36-7, holiday in Rome 33-4, map of journeys 26; second journey 32-3, the Nondescript 37-8, third journey 35, views on taxidermy 33-5, 39, Walton Hall 27, 32-3, 38-9
Waterton Lakes National Park 36
wattled honeyeaters 123
weka 124-6
West Indies 27; Antigua 37; Barbados 16, 30, 58; Bermuda 59; Dominica 37; Grenada 32; Guadaloupe 37; Jamaica 16, 18, 57-8; St Thomas's 32; St Vincent 32
White, Gilbert 14
white butterflies 63
white egrets 29
Whymper, J.M. 19, 72
Wilberforce, 'Soapy Sam' 148
wildlife conservation movement 148
wild turkey C5
willow warblers 117-18
Wilson, Alexander 35, C5
wood-quail 95
wood storks 101
wourali 30, 32

Young, Alan 48-56 *passim*

Zanzibar 82, 85
zebra 86, 90; Burchell's 88; Grant's 88
Zoological Society of London 12, 23, 85-6, 90, 94, 119